Microplastics

This book introduces the growing problem of microplastic pollution in the soil and aquatic environment and its interaction with other chemical pollutants. Further, it provides a detailed review of existing analysis techniques for characterization, separation, and quantification of microplastics including their merits and demerits with possible suggestions. Additionally, the regulatory need and actions for improving the economic and quality of plastic recycling, curbing microplastic littering, and challenges of stakeholders, researchers, and recyclers are reviewed comprehensively. Priorities are identified to bridge the knowledge gaps for appropriate management of existing challenges.

Key Features of book are as follow:

1. Provides a comprehensive description of the fate and environmental impact of microplastics, along with various characterization methods
2. Overviews the interaction of microplastics with other toxic chemicals and further their transport in the environment
3. Explains how microplastics enter the environment and their effect on biota and human health
4. Analyzes existing analytical techniques for characterization of microplastics
5. Describes societal awareness related to the use of plastic and discarding

This book focuses on graduate students, researchers in environmental engineering, ecological engineering, chemical and biological engineering, plastics and material sciences/engineering, waste management, and materials science.

Microplastics
Analytical Challenges and Environmental Impacts

Edited by
Hyunjung Kim

CRC Press
Taylor & Francis Group
Boca Raton London New York

CRC Press is an imprint of the
Taylor & Francis Group, an **informa** business

First edition published 2023
by CRC Press
6000 Broken Sound Parkway NW, Suite 300, Boca Raton, FL 33487-2742

and by CRC Press
4 Park Square, Milton Park, Abingdon, Oxon, OX14 4RN

CRC Press is an imprint of Taylor & Francis Group, LLC

Library of Congress Cataloging-in-Publication Data
CIP data has been applied for

ISBN: 9781032060774 (hbk)
ISBN: 9781032060781 (pbk)
ISBN: 9781003200628 (ebk)

DOI: 10.1201/9781003200628

Typeset in Times
by codeMantra

Microplastics
Analytical Challenges and
Environmental Impacts

Edited by
Hyunjung Kim

CRC Press
Taylor & Francis Group
Boca Raton London New York

CRC Press is an imprint of the
Taylor & Francis Group, an **informa** business

First edition published 2023
by CRC Press
6000 Broken Sound Parkway NW, Suite 300, Boca Raton, FL 33487-2742

and by CRC Press
4 Park Square, Milton Park, Abingdon, Oxon, OX14 4RN
CRC Press is an imprint of Taylor & Francis Group, LLC

Library of Congress Cataloging-in-Publication Data
CIP data has been applied for

ISBN: 9781032060774 (hbk)
ISBN: 9781032060781 (pbk)
ISBN: 9781003200628 (ebk)

DOI: 10.1201/9781003200628

Typeset in Times
by codeMantra

This book is dedicated to my beloved parents whose footprints of grace have kept me going further in life and are continuously enlightening to me.

Hyunjung Kim

Contents

Preface

Plastic is an organic polymer synthesized from fossil feedstocks such as natural gas, oil, or coal. In modern times, the first plastic was manufactured in 1907 and was called "Bakelite". Due to the many benefits of plastics such as being cheap, versatile, lightweight, and resistant, its worldwide production which was only 0.35 million metric tons in 1950 has increased nearly 200-fold, reaching 348 million metric tons in 2017, with average annual growth of 9%. After the completion of the life span, the macroplastic debris in the environment undergoes degradation by natural or anthropogenic means to form tiny plastic fragments (often between the range of 1 μm–5 mm) called microplastics.

One of the primary environmental risks associated with microplastics is that they are ubiquitous and bioavailable for injection by marine organisms, soil organisms, and plants growing on microplastic polluted soils. Furthermore, microplastics have been reported to be in tandem with other toxic chemicals serving as a vector for their transport in the environment such as heavy metals and persistent organic pollutants (e.g., polycyclic aromatic hydrocarbons, organochlorine pesticides, and polychlorinated biphenyls).

To adequately elucidate the associated risks of microplastics, identification of a full range of microplastics in the media of concern is an absolute necessity but a daunting challenge. While a standardized method for the detection of microplastics is optimal, a single procedure will not suffice for all applications due to the diversity of compositions, sizes, and shapes of microplastics, as well as interferences associated with different environmental matrices such as sediments, wastewater, sludge, suspended particulate matter, dust, and tissue. This requires the development, refinement, and application of a suite of novel integrated analytical approaches.

In the current scenario, this book provides in-depth knowledge of the synthetic chemistry of plastic, the behavior of mesoplastics, microplastics, and/or neoplastics in the environment, and its interaction with other toxic chemicals as well as associated analytical challenges.

Chapter 1 provides deep insight into the qualitative characteristics of plastics while describing their chemical nature in simple terms and the chemistry of various synthetic routes in detail. Furthermore, multiple types of classifications of material regarding plastics according to their chemical structure, degree of cross-linking, types of base material, and their applications are also discussed comprehensively in this chapter. Associated analytical and environmental challenges are also overviewed. Chapter 2 provides an overview of countries' progress in passing laws and regulations that limit the manufacture, import, sale, use, and disposal of selected macro-, meso-, micro-, and nanoplastics with environmental impacts. Chapter 3 highlights that the potential degradation pathways and products depend on the polymer type. Chapter 4 comprehensively reviews the interactions of inorganic and organic pollutants with microplastics. Chapter 5 describes the impact of microplastics and nanoplastics on soil and plant characteristics. Furthermore, the impact of microplastics and nanoplastics on invertebrates and vertebrates is also highlighted.

Chapter 6 describes various sampling strategies and devices to collect microplastics and nanoplastics from various environments. Challenges associated with sampling are also discussed in the chapter. Chapter 7 introduces various techniques for separation of microplastics from aquatic, atmospheric, and sediment-containing environments. Chapter 8 reviews the potential of scanning electron microscopy, pyrolysis–gas chromatography–mass spectrometry, Fourier-transform infrared spectroscopy, Raman spectroscopy, nuclear magnetic resonance spectroscopy, and Matrix-assisted laser desorption/ionization–time-of-flight mass spectrometry as an analytical tool for identification of microplastics and nanoplastics. The analytical challenges associated with each technique for the quantification of microplastics are also discussed in detail

We believe that this current book can be an excellent source of reference and information on microplastics for scientists, engineers, students, industry stakeholders, government sector/policymakers, and citizens.

Prof. Hyunjung Kim
Seoul, South Korea

Contributors

Humma Akram Cheema
Department of Chemistry
University of Agriculture Faisalabad
 (UAF)
Faisalabad, Pakistan

Allan Gomez-Flores
Department of Earth Resources &
 Environmental Engineering
Hanyang University
Seoul, Republic of Korea

Gilsang Hong
Department of Earth
 Resources & Environmental
 Engineering
Hanyang University
Seoul, Republic of Korea

Gukhwa Hwang
Department of Mineral Resources and
 Energy Engineering
Jeonbuk National University
Jeonju, Republic of Korea

Sadia Ilyas
Department of Earth Resources &
 Environmental Engineering
Hanyang University
Seoul, Republic of Korea

Geunbae Kim
Risk Assessment Division
National Institute of Environmental
 Research
Incheon, Republic of Korea

Hyunjung Kim
Department of Earth Resources &
 Environmental Engineering
Hanyang University
Seoul, Republic of Korea

Byoung-cheun Lee
Risk Assessment Division
National Institute of Environmental
 Research
Incheon, Republic of Korea

Editor

Hyunjung "Nick" Kim is a Professor at the Department of Earth Resources & Environmental Engineering, Hanyang University, Seoul, Republic of Korea. Dr. Kim received his B.S. and M.S. degrees from Hanyang University in Seoul, Korea, and Ph.D. from California, Riverside, in Chemical and Environmental Engineering. He served over twelve years in the Department of Mineral Resource and Energy Engineering of Jeonbuk National University, Jeonju. He specializes in colloidal chemistry, froth flotation, bio-hydrometallurgy, and microplastics with over ten years of academic research experience. Prof. Kim is currently serving as the Editorial Member/Associate Editor in several reputed journals. Prof. Kim has authored more than 150 refereed journal publications, several book chapters, and edited books. He has been awarded the Young Scientist award from the Industrial Minerals & Aggregates Division of the Society for Mining, Metallurgy, and Exploration (SME) in 2016.

1 Emergence, Chemical Nature, Classification, Environmental Impact, and Analytical Challenges of Various Plastics

Hyunjung Kim and Sadia Ilyas
Hanyang University

CONTENTS

DOI: 10.1201/9781003200628-1

"Plastic" is an umbrella term encompassing a wide range of materials made of semi-synthetic or synthetic organic compounds. The International Union of Pure and Applied Chemistry defines plastics as polymeric materials that may contain other substances to improve performance and/or reduce costs" (Vert et al., 2012). The main feature of these materials is reflected in their etymology: the word plastic originates from the Greek words plastikos (a Greek term), meaning "capable of being shaped," and plastos (Latin term), meaning "molded." Typically synthetic, plastics are most commonly derived from petrochemicals and exhibit high molecular mass and plasticity. Given the prevalence of plastics in our society, it is not surprising that some researchers have called our present-day "the Plastic Age," but named age should not be confused with geological divisions of time, such as the present Holocene (11,650 years ago – present) or the proposed Anthropocene, a geological era characterized by humans as a geological force and process, ushered in by the nuclear age and perpetuated by plastics (Crawford and Quinn, 2017; da Costa et al., 2016).

The incredible versatility of this group of materials accounts for the continued growth in production year after year. In tandem with that growth, the market value of plastics also continues to grow. The global plastic market size increased from approximately 502 billion U.S. dollars in 2016 to 579.7 billion U.S. dollars in 2020. It is forecast that the worldwide plastic market will grow to a value of 750.1 billion U.S. dollars by 2028.[1]

The current long-term forecasts suggest that the production of plastic materials shows no sign of leveling off and is expected to surge exponentially (Ng and Obbard, 2006). Thus, by 2050 it is anticipated that around an extra 33 billion tons of plastic will have been produced, and annual global production shall be between 850 million tons and 1,124 million tons. Nevertheless, it is challenging to predict this far into the future with any degree of certainty (Rochman et al., 2013; Ng and Obbard, 2006).

In 2019, China had a total share of approximately 31% of the global production of plastic materials, which made it the world's largest plastic producer. NAFTA was the second-largest plastic producing region after China, accounting for 19% of global production.[2]

Generally, a piece of plastic equal to or larger than 25 mm in size is termed a macroplastic. Additionally, pieces of plastic smaller than 25 mm to 5 mm in size along its longest dimensions are categorized as mesoplastics. While pieces of plastic smaller than 5 mm to 1 mm in size along its longest dimensions are considered microplastics and less than 1 μm as nanoplastics, the size of nanoplastics is still debatable (da Costa et al., 2016). Microplastics are of two types, namely primary and secondary microplastics. Primary microplastics are typically small spherical microbeads intentionally manufactured by the plastics industry for use in cosmetics, personal care products, dermal exfoliators, cleaning agents, and sandblasting shots (Crawford and Quinn, 2017).

The synthetic fibers used to produce garments are also considered to be primary microplastics since they were purposely manufactured to be of a small size.

---

[1] https://www.statista.com/statistics/1060583/global-market-value-of-plastic/.
[2] https://www.statista.com/statistics/281126/global-plastics-production-share-of-various-countries-and-regions/.

In samples recovered from the environment, fibers are among one of the most abundantly found plastic pieces. One particularly significant source of microplastic fibers is from the washing of fabrics in industrial settings and by consumers. Furthermore, manufacturers of cleaning products used on human skin incorporate microbeads to aid in deep cleansing and dermal abrasion of the skin. Microbeads are often added to toothpaste for whitening purposes and even deodorants to block pores and prevent perspiration (Crawford and Quinn, 2017).

Aside from fibers, wastewater often contains large amounts of other plasticles, such as microbeads (small spherical particles of plastic less than 1 mm to 1 μm in size) and nanoplastics. Manufacturers of cosmetics often incorporate these tiny plasticles into many of their products to facilitate control over the viscidness of the product and for the creation of films (Crawford and Quinn, 2017; da Costa, 2018).

Secondary microplastics result from the breakdown of larger plastic particles. Over time, a combination of physical, chemical, and biological processes can reduce the structural integrity of these plastics, leading to their fragmentation (da Costa, 2018). However, this breakdown can also take place before these materials enter the environment, as is the case of synthetic fibers from clothes released during washing cycles or the wear-and-tear of car tires, which generates minute polymeric fragments (Sommer et al., 2018).

Smaller microplastics, called nanoplastics, can also be present in the environment (da Costa et al., 2016; Gigault et al., 2018). Similar to microplastics, nanoplastics may be released into the environment directly or due to the fragmentation of larger particles. Hence, nanoplastics may also be classified as primary or secondary nanoplastics. Primary nanoplastics include particles found in paints, adhesives, and electronics. Also, thermal cutting of polystyrene or polyvinylchloride (PVC), and 3D printing result in the release of particles as small as 11.5 nm into the environment. Secondary nanoplastics form from the fragmentation of larger plastic particles, such as microplastics. However, the exact mechanism through which this occurs is still unclear (Zhang et al., 2012; Zhan et al., 2017; Stephens et al., 2013).

1.1 A HISTORICAL PERSPECTIVE OF PLASTICS

The advent of plastics is quite indistinct, that can contend, to begin with, the creation of rubber balls by the Mesoamericans around 1600 BC and later by the introduction of natural rubber by French explorer Charles Marie de La Condamine in 1736 from the Pará rubber tree (*Hevea brasiliensis*) (Millet et al., 2018).

In 1843, Thomas Hancock, a British manufacturing engineer, patented the vulcanization of natural rubber in the United Kingdom, and American chemist Charles Goodyear patented the vulcanization of natural rubber in the United States 8 weeks later in 1844. The original breakthrough for the first semi-synthetic plastics material, cellulose nitrate, occurred in the late 1850s and involved the modification of cellulose fibers with nitric acid. Cellulose nitrate had many false starts and financial failures until a Briton, Alexander Parkes, exhibited the so-called "Parkesine" as the first world's synthesized plastic in 1862. However, due to its high manufacturing costs, the failure of this product led to the creation of Xylonite by Daniel Spill. This new material started finding success in producing objects such as ornaments, knife handles, boxes, and more flexible products such as cuffs and collars. In 1869, an American, John W. Hyatt, made a revolutionary discovery, a process to produce celluloid. This product could be used to substitute for natural substances such as tortoiseshell, horn, linen, and ivory. This product entered mass production in 1872.[3]

Until the early 1900s, it was impossible to use cellulose nitrate at very high temperatures, because it was flammable. The development of cellulose acetate brought about a solution to this problem. It started being used as a non-flammable "dope" to stiffen and waterproof the fabric wings and fuselage of early airplanes and was later widely used as a cinematographic "safety film." In the meantime, casein formaldehyde was developed based on fat-free milk and rennin and used for shaping buttons, buckles, and knitting needles.[4, 5]

The Discovery of Bakelite. In 1907, Belgian Leo Baekeland (who coined the term plastic later on) discovered Bakelite, which was largely used in the expanding automobile and radio industries at that time.[6]

In 1912, a German chemist, Fritz Klatte, discovered PVC and polyvinyl acetate (PVA). The following year, Jacques E. Brandenberger, a Swiss engineer, invented Cellophane, a transparent, flexible, waterproof packaging material.

In 1921, the first injection molding press appeared, invented by Arthur Eichengrün. Meanwhile, a revolution came in 1922, when a German, Herman Staudinger, claimed molecules could join to form long chains and become "macromolecules" or polymers. Staudinger provided enough evidence for his macromolecular concept and promoted it, despite the strong opposition of several chemists. Staudinger provided the theoretical basis for polymer chemistry and significantly contributed to the rapid development of the polymer and plastic industry and was awarded Nobel Prize for chemistry in 1953.[7]

In 1927, another important scientific breakthrough occurred, when Waldo Semon, an American researcher, found a way to plasticize PVC, which had been discovered more than a decade before. PVC was thus converted into a flexible material that could be used for flooring, electrical insulations, and roofing membranes (Millet et al., 2018).

In 1930, the commercial production of polystyrene started. In the meantime, Otto Röhm invented a great product in 1933, Plexiglas™, "a crystal-clear, shatter-proof polymethyl acrylate sheet" which found an important market in the aircraft industry.

In 1935, Wallace Carothers from DuPont Company was the first to synthesize Nylon™ (polyamide), which became very famous in stockings. The first commercial PVC products were introduced in 1934 and 1935. These were flooring and pipes, respectively. Three years later, a Swiss researcher, Pierre Castan, patented the synthesis of epoxide resins, initially used in dentistry (for dental fixtures and castings), as well as medicine. Their properties were also valuable as a constituent of glue (Chandler, 2005).

---

[3] http://www.robinsonlibrary.com/technology/chemical/biography/spill.htm.
[4] http://www.robinsonlibrary.com/technology/chemical/biography/spill.htm.
[5] https://www.plasticseurope.org/application.
[6] https://www.chemheritage.org/the-history-and-future-of-plastics.
[7] https://www.acs.org/content/dam/acsorg/education/whatischemistry/landmarks/staudinger polymerscience.

In the 1940s, World War II meant a boost for plastics production and further development, which took on a key role in the military supply chain. Plastics were used to make almost everything: for example, nylon could be found in parachutes, ropes, body armor, and helmet liners, while Plexiglas™ replaced glass in aircraft windows. A wide variety of pioneering materials were invented during the wartime period, such as polyethylene, polystyrene, polyester, polyethylene terephthalate, silicones, and many more (Millet et al., 2018).

In the early 1950s, Hogan and Bank discovered that ethylene can be polymerized under moderate conditions (3–4 MPa and 70°C–100°C) with chromium oxide catalysts on silica support. This procedure's manufacture of high-density polyethylene (HDPE) started in 1956. Currently, HDPE is one of the largest commodities of plastic. Shortly after (1953), Ziegler and Natta independently developed a family of stereospecific transition-metal catalysts that led beside HDPE to the synthesis and commercialization in 1957 of polypropylene (PP) as a major commodity plastic. Additionally, in the 1950s, the growth of plastics for domestic use was spread out. Decorative laminates were invented, such as Formica™ tables, which were very popular in the US and used in espresso bars and diners. In the same period, plastics also became a major force in the clothing industry. Polyester, Nylon™, and Lycra™ fabrics were easy to wash, needed no ironing, and often were cheaper than their natural alternatives (Krentsel et al., 1997; Kresser, 1961).

In the 1950s, polycarbonate-related production commenced. Initially, in 1898 Einhorn reported the formation of polycarbonates (PC) in the reaction of phosgene with dihydric phenols. In 1902 Bishoff and Hedenstrom confirmed this work using an ester-exchange reaction. These PC were intractable, insoluble materials. For many years the plastics industry showed no interest in their development, though admittedly, Carothers included aliphatic polycarbonates in his polyesters and reported on these in 1930. That same year Carothers and Natta prepared several aliphatic PC. When bisphenol A was readily available commercially due to contributions made by Schnell in 1956, the time was opportune for the development of bisphenol-A polycarbonates. Such polymers became available for assessment in 1958, and full-scale production commenced in 1959. In the same period, styrene-acrylonitrile copolymer (1954) and polyacetals (1956) were synthesized for the first time (Brydson, 1989; Goodman and Rhys, 1965).

Polyimide (PI) resins, primarily of the thermosetting type, were developed after 1953 and commercially available around 1963. The thermoplastic PI result from the polyaddition of bismaleinimides and HZ-R-ZH compounds, where HZ can be diamine, disulfide, dialdoxim, etc. Although aromatic PI retains usable properties at 300°C for months and withstands exposures of a few minutes well over 500°C, their applications have been limited. The 1960s are also known as a decade of mass distribution of stylish, innovative, and impressive plastic products in the fashion world, such as soft and hard foams with protective skin, wet-look polyurethane, transparent acrylic, and artificial leather. The home decoration was also enriched, where unconventional designer furniture such as inflatable chairs and acrylic lights became important for fashion-conscious consumers. Moreover, plastic materials played an essential role in the production of spacecraft components; their lightness and versatility made them irreplaceable for the success of space exploration (Chandler, 2005).

In 1974, liquid crystal behavior for a polymeric system was observed in a concentrated solution of poly(benzyl L-glutamate) which forms a rod-like α-helical conformation in various organic solvents. Liquid crystal polyesters may have been recognized as early as 1965; they have highly aromatic structures. Polysulfone was introduced around 1965. Polyarylsulfones were commercialized in the period 1967–1976 and polyethersulfone around 1972. All polysulfones are transparent, light yellow to amber polymers. Their high heat distortion temperatures are only surpassed by aramids, polyimides, and polyamide imides. In 1976 Kaminsky and Sinn discovered a new family of catalysts (a metallocene complex) for ethylene polymerization. Kaminsky–Sinn catalysts and various modifications afforded the synthesis of ethylene-α olefin copolymers with a high degree of compositional uniformity. Polyether ether ketone, the most important member of the poly(aryl ether ketone)s, became commercially available in 1978 (Fried, 1995; Elias, 1993).

In the 1980s, the rise of global communications directly impacted the production and use of plastics, which provided the raw material for the production of personal computers, fiber optic cables, and portable telephones. In transport, the demand for plastics in cars also increased. In the 1980s, the first flight tests of an all-plastic aircraft took place. Moreover, plastic packaging became very important in shopping because it helped distribute and preserve the quality of the products we buy from supermarkets.

In the 1990s and 2000s, plastics became key components for meeting challenging societal demands. Plastics are currently essential in designing structural elements such as insulation, life support systems, space-suit fabric, food packaging, guidance and communication systems, solar panels, and so forth. The consumer demands for longer product shelf lives and freshness retention led to the development of plastic packaging with superior barrier properties. Raised awareness in society of the necessity to save fossil fuels increased the need for plastic products, enabling improvement in the energy efficiency of buildings and a reduction in fuel consumption in transportation[8] (Chandler, 2005).

1.2 CHEMICAL NATURE AND SYNTHESIS ROUTES OF PLASTICS

It is quite challenging to produce an adequate definition of the word "plastics," all plastics materials, before compounding with additives, consist of a mass of very large molecules (macromolecules), which, in turn, are composed of many repeating small pieces molecules called monomers. The chemical reaction in which macromolecules are formed from monomers is known as polymerization. There are two main types of polymerization, namely chain reaction polymerization and step reaction polymerization, often referred to by their older names, addition, and condensation polymerization (Billmeyer, 1984; Young and Lovell, 2011).

1.2.1 THE CHEMISTRY OF CHAIN REACTION POLYMERIZATION

In chain reaction polymerization, the individual monomers are added to one another in sequence to form a long chain. In a typical example of chain reaction

---

[8] https://www.bbc.com/news/magazine-27442625.

polymerization, a simple, low molar mass molecule possessing a C=C double bond (monomer) reacts with an initiator or already-growing polymer chain by breaking the double bond, leaving a free valence for further reaction with other monomers. Several chemical mechanisms can be used to polymerize C=C bonds, such as free radical polymerization, controlled radical polymerization, ionic polymerization, and coordination polymerization, but all have in common three main stages: initiation, propagation, and termination (Crawford and Quinn, 2017; Edmondson and Gilbert, 2017; Gregg and Mayo, 1947).

1.2.1.1 Chemistry of Free-Radical Polymerization

In free-radical polymerization, the initiation stage begins with generating a radical containing reactive unpaired electrons. The radical is typically generated from the decomposition of an added initiator triggered either by warming, irradiation with ultraviolet light, or reaction with other chemicals. Two common initiators which decompose on warming are azobisisobutyronitrile and benzoyl peroxide.

Such free radicals are generated as:

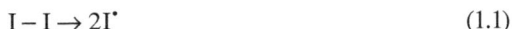

$$I-I \rightarrow 2I^{\cdot} \tag{1.1}$$

The rate of formation of radicals depends on the concentration of the initiator, temperature, and the nature of other molecules present. In most polymerization reactions, this stage requires a conversion time of at least 30 min, while all subsequent stages of polymer growth occur almost instantaneously.

The above generated radical (Eq. 1.1) can react with another monomer molecule by addition reaction to produce a second radical as in Eq. (1.2).

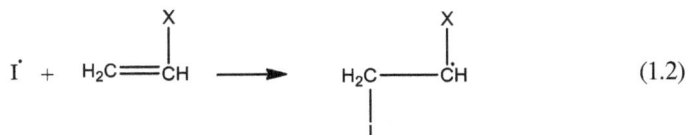

$$I^{\cdot} + H_2C{=}CH{-}X \longrightarrow H_2C{-}\overset{X}{\underset{|}{CH}} \tag{1.2}$$

The second generated radical can react further with monomer molecules to produce another radical of similar reactivity (Eq. 1.3)

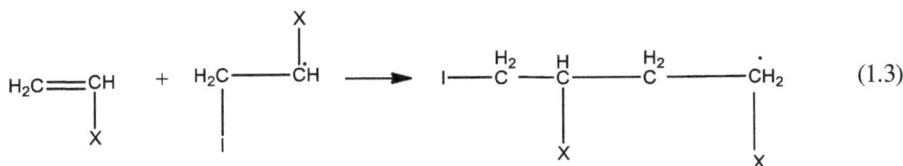

$$H_2C{=}CH{-}X + H_2C{-}\overset{X}{\underset{|}{CH}} \longrightarrow I{-}\overset{H_2}{C}{-}\overset{H}{\underset{X}{C}}{-}\overset{H_2}{C}{-}\overset{\cdot}{CH_2} \tag{1.3}$$

This reaction may repeatedly repeat itself to unite several thousand monomer units in time order of 1 s, leading to the propagation stage. After propagation, the chain termination can occur in various ways, including a mutual combination of two growing radicals, disproportionation between growing radicals, reaction with an initiator radical, and chain transfer with a modifier (Eqs. 1.4–1.7).

$$\text{wwH}_2\text{C}\text{---}\overset{\cdot}{\text{C}}\text{H} \quad + \quad \text{H}\overset{\cdot}{\text{C}}\text{---}\overset{\xi}{\text{CH}_2} \quad \longrightarrow \quad \text{wwH}_2\text{C}\text{---}\overset{\text{H}}{\underset{\text{X}}{\text{C}}}\text{---}\overset{\text{H}}{\underset{\text{X}}{\text{C}}}\text{---}\underset{\text{H}_2}{\text{C}}\text{wwww} \tag{1.4}$$

$$\text{wwH}_2\text{C}\text{---}\overset{\cdot}{\text{C}}\text{H} \quad + \quad \text{H}\overset{\cdot}{\text{C}}\text{---}\overset{\xi}{\text{CH}_2} \quad \longrightarrow \quad \text{wwHC}\text{==}\underset{\text{X}}{\text{CH}} \quad + \quad \text{H}\overset{\cdot}{\underset{\text{X}}{\text{C}}}\text{---}\underset{\text{H}_2}{\text{C}}\text{wwww} \tag{1.5}$$

$$\text{wwH}_2\text{C}\text{---}\overset{\cdot}{\underset{\text{X}}{\text{C}}}\text{H} \quad + \quad \text{I}^{\cdot} \quad \longrightarrow \quad \text{wwwCH}_2\text{---}\overset{\text{H}}{\underset{\text{X}}{\text{C}}}\text{---I} \tag{1.6}$$

$$\text{wwH}_2\text{C}\text{---}\overset{\cdot}{\underset{\text{X}}{\text{C}}}\text{H} \quad + \quad \text{RY} \quad \longrightarrow \quad \text{wwwCH}_2\text{---}\overset{\text{H}}{\underset{\text{X}}{\text{C}}}\text{---Y} \quad + \quad \overset{\cdot}{\text{R}} \tag{1.7}$$

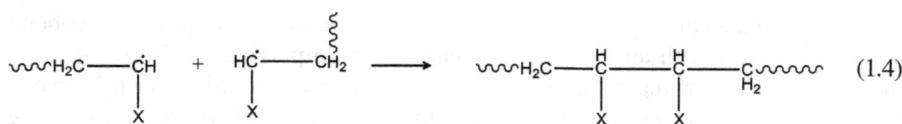

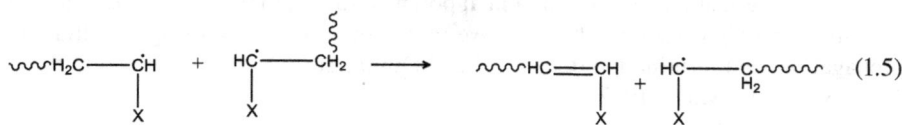

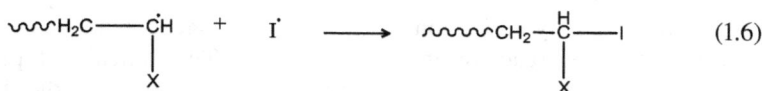

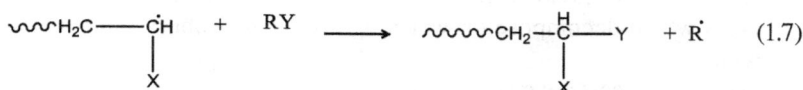

Mainly termination occurs without affecting the reaction rate and with no net loss in radical concentration. Additionally, chain transfer with polymer or monomer by abstraction of an H atom (branched polymers) and reaction with a molecule to form stable free radical, as in the case of hydroquinone, can also occur (Matyjaszewski and Davis, 2003; Tsarevsky and Sumerlin, 2013; Baysal and Tobolsky, 1952).

1.2.1.2   Chemistry of Ionic Polymerization

Chain reaction polymerizations produce several important polymeric plastics with ionic mechanisms. Although the process involves initiation, propagation, and termination stages, the chain growing unit is an ion rather than a radical. The electron distribution around the carbon atom of a growing chain may further take several forms, such as free radical, carbocation, and carbanion (as depicted in Figure 1.1).

Both carbocations and carbanions can be used as the active centers for chain growth in cationic and anionic polymerization. These polymerizations often occur at a very high rate, and traces of co-catalysts can significantly affect the reaction rate. Monomers with electron-donating substituents tend to form carbocations, while those with electron-withdrawing substituents can be polymerized by anionic polymerization (form carbanion). Free-radical polymerization is somewhat intermediate and is possible when substituents have moderate electron-withdrawing characteristics. Many monomers may be polymerized by more than one mechanism (Edmondson and Gilbert, 2017).

Cationic polymerization is used commercially for the synthesis of poly-formaldehyde and poly-isobutylene that is catalyzed by Friedel-Crafts agents such as aluminum chloride ($AlCl_3$), titanium tetrachloride ($TiCl_4$), and boron trifluoride (BF_3), as strong electron acceptors, in the presence of a co-catalyst. High molar mass products may be obtained within a few seconds at 100°C by these polymerization reactions. The first stage involves the reaction of the catalyst with a co-catalyst (e.g., water) to produce a complex acid, as in Eq. (1.8).

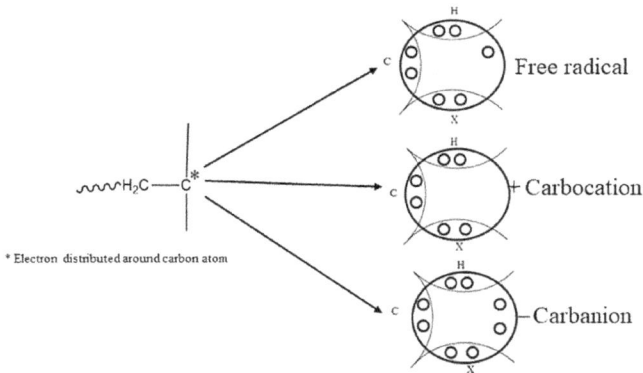

FIGURE 1.1 Various forms of electron distribution around the carbon atom of a growing chain.

$$TiCl_4 + RH \rightarrow TiCl_4R^-H^+ \tag{1.8}$$

This donates a proton to the monomer to produce a carbocation as an initiation step (Eq. 1.9). In turn, this ion reacts with a further monomer molecule to form another reactive carbocation (Eq. 1.10) as a chain propagation step.

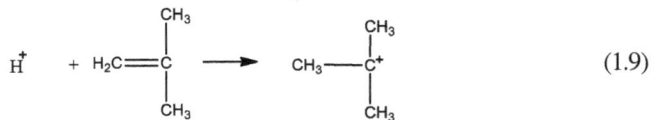

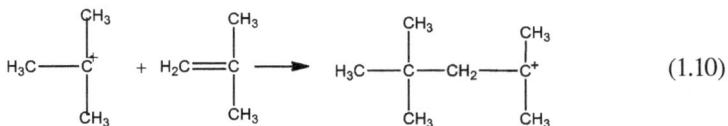

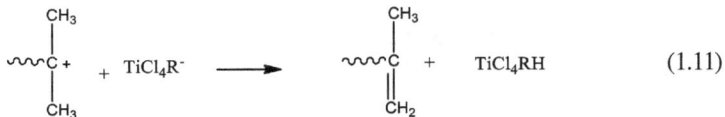

Termination can occur by rearrangement of the ion pair (Eq. 1.11) or by chain transfer to monomer.

Anionic polymerization was first used in the sodium-catalyzed production of polybutadiene (synthetic rubber). Typical catalysts include alkali metals, alkali metal alkyls, and sodium naphthalene. Although the process is of particular importance in producing synthetic rubbers instead of plastics but is indirectly co-related to improve the characteristic properties of plastics. Approximately 25% of that synthetic rubber (polybutadiene) is used as an additive to enhance the toughness of plastics such as polystyrene and acrylonitrile butadiene styrene (ABS).

Anionic polymerization does not necessarily imply the presence of a free anion on the growing polymer chain but can more likely proceed when there are electron-withdrawing substituents present in the monomer (e.g., –CN or phenyl). In principle, initiation may occur either by adding an anion to the monomer (Eq. 1.12) or by the addition of an electron to produce a radical anion, as in Eq. (1.13).

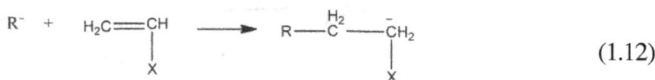

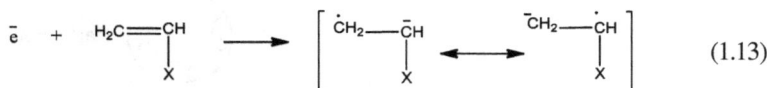

$$R^- \; + \; H_2C{=}CH{-}X \longrightarrow R{-}\overset{H_2}{C}{-}\overset{-}{C}H_2{-}X \tag{1.12}$$

$$\bar{e} \; + \; H_2C{=}CH{-}X \longrightarrow \left[\overset{\cdot}{C}H_2{-}\overset{-}{C}H{-}X \longleftrightarrow \overset{-}{C}H_2{-}\overset{\cdot}{C}H{-}X \right] \tag{1.13}$$

The most common initiators are the alkyl and aryl derivatives of alkali metals, particularly lithium. There is commonly no termination step in anionic polymerizations in the presence of impurities, and the monomer continues to grow until all are consumed. Under certain conditions, the addition of further monomer, even after several weeks, can cause the dormant polymerization process to proceed. The process is known as living polymerization and the products as living polymers. Of particular interest is that the follow-up monomer may be of a different species, enabling the production of block copolymers. This technique is essential with certain types of thermoplastic elastomer and some rather specialized styrene-based plastics (Edmondson and Gilbert, 2017).

A further feature of anionic polymerization is that, under very carefully controlled conditions, it may be possible to produce a monodisperse polymer in contrast to free-radical polymerizations that facilitate the production of polydisperse polymer samples. Prerequisites to produce monodisperse polymers include:

1. All the growing chains must be initiated simultaneously.
2. All the growing chains must have equal growth rates.
3. There must be no transfer or termination reaction so that all chains continue to grow until all monomers are consumed.

The number average degree of polymerization ($\overline{X}_n$, defined as the average number of monomer units per polymer chain) can be represented by Eq. (1.14).

$$\overline{X}_n = nx\frac{[M]}{[I]} \tag{1.14}$$

Where [M] and [I] are the initial monomer and initiator concentrations, respectively; n is equal to 1 or 2 depending on whether each initiator begins one chain (when initiating with organometallic compounds) or two (when initiating by electron addition), and x is the fraction of monomer converted into polymer. In principle, it is possible to extend the method to produce block copolymers in which each block is monodisperse, but avoiding impurities becomes challenging. Nevertheless, narrow size distributions are achievable. Another feature of anionic polymerization is the possibility of coupling chains together at their "living ends." A stable non-living linear polymer can be produced with a bi-functional coupling agent, with twice the average length of the starting molecules. Bi-functional coupling agents can also produce tri-block copolymers by coupling together di-block chains, as in one route to commercial SBS thermoplastic elastomers. Multifunctional coupling agents will result in star polymers, with several linear chains radiating from a central branch point. An example is the coupling of butyllithium-initiated polystyrene with silicon tetrachloride (Eq. 1.15) (Edmondson and Gilbert, 2017; Flory, 1953):

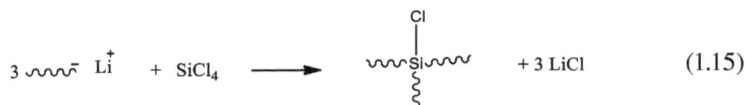

$$3 \text{ } \sim\!\sim\!\sim^{-} \text{ } \overset{+}{\text{Li}} \quad + \text{ SiCl}_4 \quad \longrightarrow \quad \sim\!\sim\!\sim\!\overset{\overset{\displaystyle \text{Cl}}{|}}{\underset{\xi}{\text{Si}}}\!\sim\!\sim\!\sim \quad + 3 \text{ LiCl} \tag{1.15}$$

1.2.1.3   Chemistry of Coordination Polymerization

Common examples of coordination polymerization are Ziegler Natta and Metallocene Polymerization after the work of Ziegler in Germany; Natta in Italy; and Pease & Roedel in the United States. The development of the Ziegler-Natta mixed organometallic catalysts in the 1950s revolutionized the manufacture of polypropylene. For the first time, coordination catalysis produces atactic polypropylene with blocks of isotactic and syndiotactic polypropylene in the chain, resulting in significantly improved properties over atactic polypropylene. Parenthetically, alterations to the conditions in which poly(methyl methacrylate) is polymerized can also impart a configuration that is either isotactic or syndiotactic. Ultimately, commercially available polypropylene is typically a mixture of tactic forms in the ratio of 75% isotactic to 25% atactic. This mixture of tactic forms imparts a material with a high molecular weight and excellent tensile strength and stiffness, although impact strength is moderate. However, manufacturers can alter the ratio to obtain the desired properties. For example, the three distinct stereochemical forms of polypropylene have different properties (Figure 1.2), as follows (Edmondson and Gilbert, 2017; Kataoka et al., 2013; Cossee, 1964):

1. Isotactic polypropylene is a highly crystalline rigid material with a high melting point of 160°C–166°C.
2. Syndiotectic polypropylene is a semicrystalline substance with elastomeric properties. 30% crystallinity imparts a melting point of 130°C.
3. Atactic polypropylene is an amorphous and tacky rubber-like material with no specific melting point.

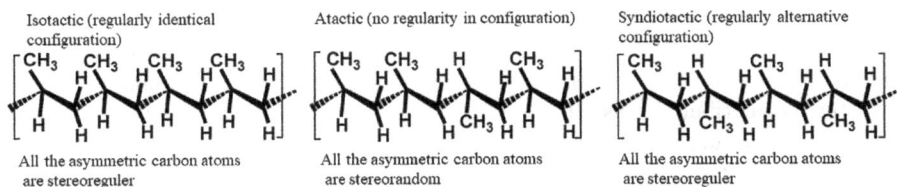

FIGURE 1.2   Various stereochemical forms of polypropylene.

More recent developments have centered on single-site catalysts such as metallocene systems. Such metallocenes typically consist of a metal atom, usually titanium or zirconium, linked with two aromatic five-carbon rings and two other groups (primarily carbon of $-CH_2$ groups). The reactivity of the catalyst can be controlled by varying the nature of substituents on the five-carbon rings, by modifying the symmetry of the structure after changing the position of substituents, and by the use of co-catalysts such as methyl aluminoxanes (Edmondson and Gilbert, 2017).

1.2.2 Chemistry of Step Reaction Polymerization

In this form of polymerization, initiation and termination stages do not exist, and chain growth occurs by random reaction between two functional groups present on the monomers or growing polymer chains. An example of a step reaction polymerization is the production of polyamide 66 (Eq. 1.16):

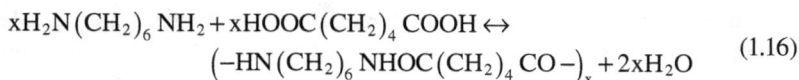

$$xH_2N(CH_2)_6 NH_2 + xHOOC(CH_2)_4 COOH \leftrightarrow$$
$$(-HN(CH_2)_6 NHOC(CH_2)_4 CO-)_x + 2xH_2O \quad (1.16)$$

Removal of water drives the reaction to the right, and a linear polymer can be formed. Polymerization proceeds in separate steps (Matyjaszewski and Davis, 2003; Edmondson and Gilbert, 2017).

1.3 TYPES OR CLASSIFICATION OF PLASTICS

There are various types of classifications regarding plastics according to their chemical structure, degree of cross-linking, types of base material, and their applications.

1.3.1 Classification Based on Chemical Structure, Crosslinking, and/or Base Material

Based on chemical structure and the degree of crosslinking, plastics are distinguished into different classes as thermoplastics, elastomers, and thermosets (Frank and Biederbick, 1984; Braun, 1985). While compounds like polymer blends, copolymers, and composite materials are composed of several base materials. This composition can be done on a physical basis (e.g., polymer blends or composite materials) or on a chemical basis (copolymers) (Klein, 2011), as indicated in Figures 1.3 and 1.4.

1.3.1.1 Thermoplastic

Thermoplastics consist of macromolecular chains with no crosslinks between the chains. The macromolecular chains can have statistically oriented side chains or build statistically distributed crystalline phases. The chemistry and structure of thermoplastic resins influence the chemical resistance and resistance against environmental effects like UV radiation. Thermoplastic resins can vary from optically transparent to opaque, depending on the type and structure of the material. Thermoplastic resins can be reversibly melted by heating and resolidified by cooling without significantly

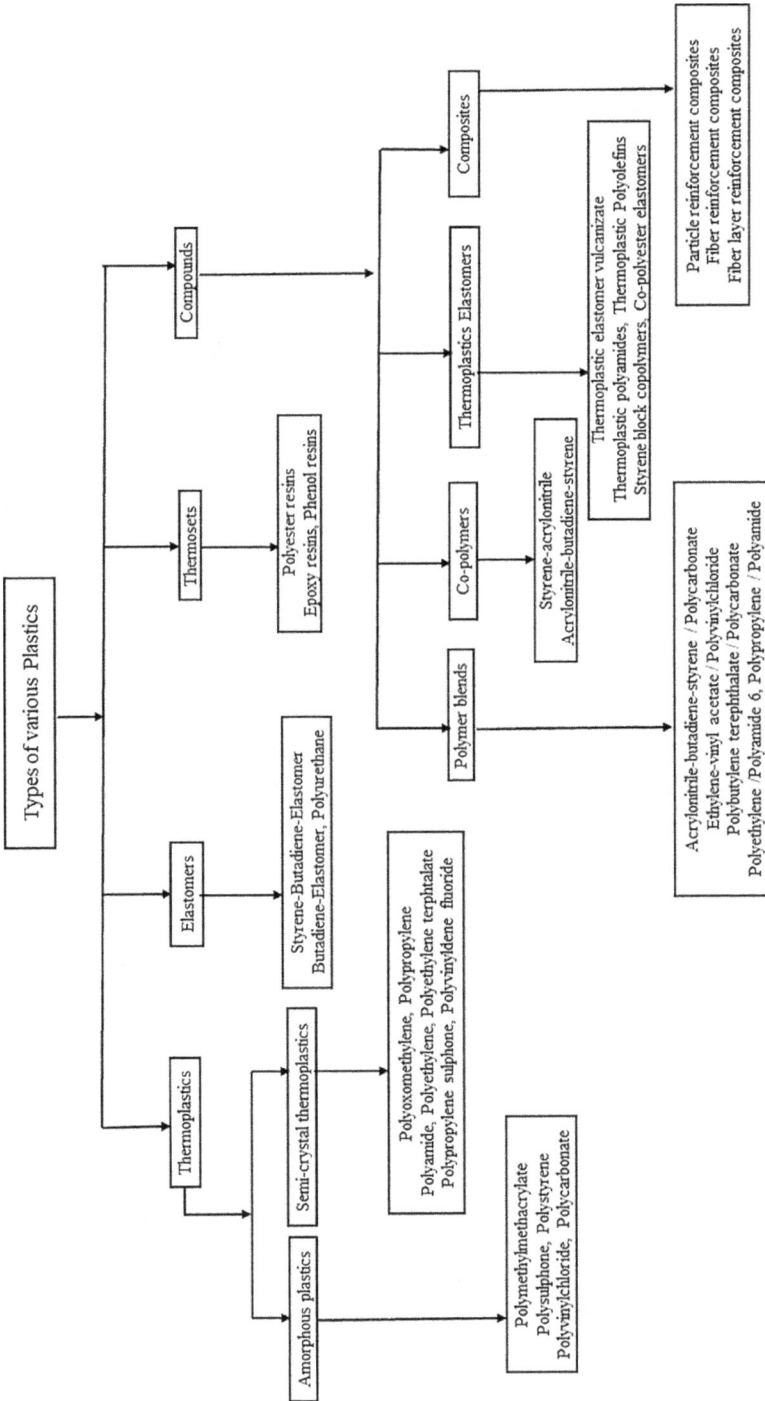

FIGURE 1.3 Schematics of various types of plastics based on structure, crosslinking, and base material. (Modified from Klein, 2011; Edmondson and Gilbert, 2017)

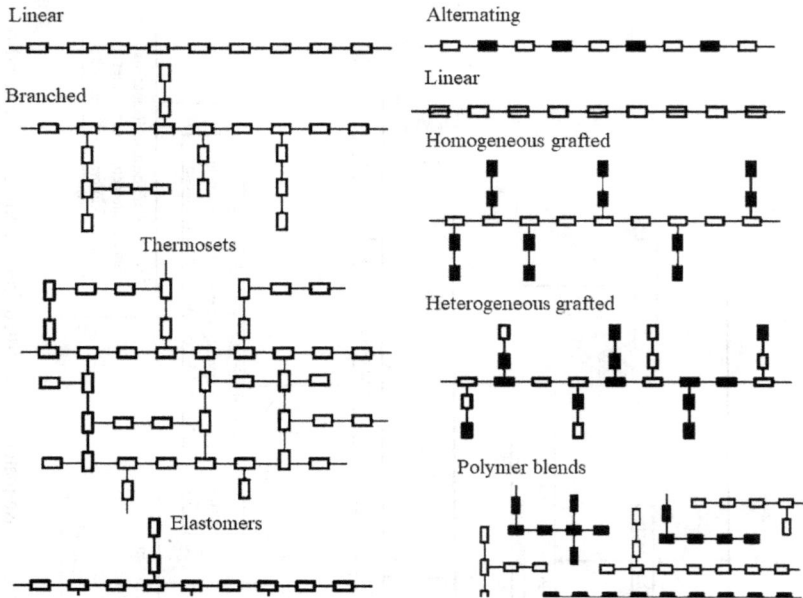

FIGURE 1.4 Principle structure of thermoplastics (linear, branched), thermosets, elastomers, copolymers (alternating, linear, homogeneous grafted, heterogeneous grafted), and polymer blends. (Modified from Edmondson and Gilbert, 2017; Crawford and Quinn, 2017; Klein, 2011.)

changing mechanical and optical properties. The melt's viscosity depends on the inner structure, like average molecular weight and spreading of the molecular weight around the average value (Skrovanek et al., 1986). The macromolecular structure of thermoplastics can be given by the chemical structure of the monomer units, the order of the monomer units in the molecule chain, and the existing side chains. Therefore, thermoplastic resins can be amorphous and semicrystalline types (Klein, 2011; Batzer, 1985; Frank and Biederbick, 1984). Figure 1.5 indicates amorphous and semicrystalline thermoplastics with typical material properties and temperature behavior.

Amorphous thermoplastic resins consist of statistically oriented macromolecules without any near order. Such resins are, in general, optically transparent and mostly brittle. Typical amorphous thermoplastic resins are PC, polymethylmethacrylate (PMMA), polystyrene, or PVC. The temperature state for applying amorphous thermoplastic resins is the so-called glass condition below the glass temperature T_g. The molecular structure is frozen in a definite shape, and the mechanical properties are barely flexible and brittle. On exceeding the glass temperature, the mechanical strength will decrease by increased molecular mobility, and the resin will become soft elastic. On reaching the flow temperature T_f, the resin will come into the molten phase. Within the molten phase, the decomposition of the molecular structure begins by reaching the decomposition temperature T_d.

Resin	Temperature of use [°C]	Specific weight [g/cm³]	Tensile strength [N/mm²]
PC	−40–120	1.2	65–70
PMMA	−40–90	1.18	70–76
PS	−20–70	1.05	40–65
PSU	−100–160	1.25	70–80
PVC	−15–60	1.38–1.24	40–60

Resin	Temperature of use [°C]	Crystallization grade [%]	Specific weight [g/cm³]	Tensile strength [N/mm²]
PA 6	−40–100	20–45	1.12–1.15	38–70
HDPE	−50–90	65–80	0.95–0.97	19–39
PETP	−40–110	0–40	1.33–1.38	37–80
PP	−5–100	55–70	0.90–0.91	21–37
PPS	<230	30–60	1.35	65–85
PVDF	−30–150	~52	1.77	30–50

Amorphous thermoplastics

Semi-crystalline thermoplastics

Elongation (%)

Tensile strength (N/mm²)

Temperature (°C)

hard elastic, brittle glass phase | T_g | soft elastic | T_f | plast. | T_d

brittle | T_g | tough elastic, hard | T_m | malleable | T_d

state of application

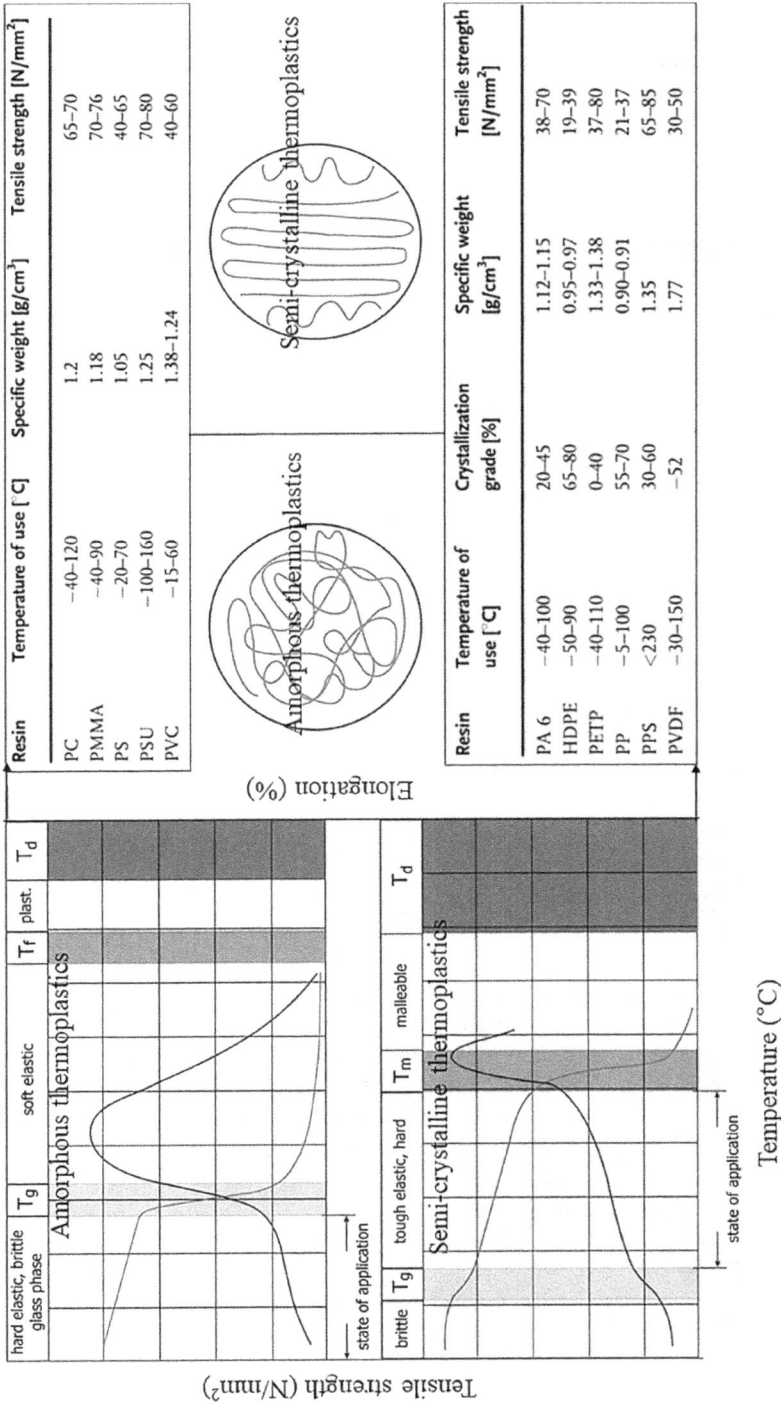

FIGURE 1.5 Amorphous and semicrystalline thermoplastics with typical material properties and temperature behavior. (Derived from Klein, 2011; Batzer, 1985; Frank and Biederbick, 1984.)

Semicrystalline thermoplastic resins consist of statistically oriented macromolecule chains as an amorphous phase with embedded crystalline phases built by near-order forces. Such resins are usually opaque and tough elastic. Typical semicrystalline thermoplastic resins are polyamide, PP, or B, polyoxymethylene (POM). The crystallization grade of semicrystalline thermoplastic depends on the regularity of the chain structure, the molecular weight, and the mobility of the molecule chains, which can be hindered by loop formation (Batzer, 1985). Due to the statistical chain structure of plastics, complete crystallization is not feasible on a technical scale. Maximum technical crystallization grades are of the order of approximately 80% (Figure 1.5).

The processing conditions can control the process of crystallization. Quick cooling of the melt hinders crystallization. Slowly cooling or tempering at the crystallization temperature will increase the crystallization grade. Semicrystalline thermoplastics with low crystallization grade and small crystallite phases will be more optically transparent than materials of high crystallization grade and large crystallite phases. Below the glass temperature T_g the amorphous phase of semicrystalline thermoplastics is frozen, and the material is brittle. Above the glass temperature, usually, the application's state (Frank and Biederbick, 1984), the amorphous phase thaws, and the macromolecules of the amorphous phase gain more mobility. The crystalline phase still exists, and the mechanical behavior of the material is tough elastic to hard. Above the crystal-melt temperature T_m the crystalline phase also melts, and the material becomes malleable. As for amorphous thermoplastics, the flowability of semicrystalline thermoplastics in the molten phase is characterized by the melt-flow index. The melt temperature of semicrystalline thermoplastics depends, among other things, on the size of the crystallites and the ratio between the amorphous and crystalline phases. A larger size and a higher proportion of crystallites will increase the melt temperature (Avramova and Fakirov, 1981). As with amorphous thermoplastics, degradation of semicrystalline thermoplastics will start in the molten phase by exceeding the decomposition temperature T_d.

1.3.1.2 Elastomers

Elastomers are plastics with wide netlike crosslinking between the molecules. Usually, they cannot be melted without degradation of the molecule structure. Above the glass temperature T_g, as the state of the application, elastomers are soft elastic. Below T_g they are hard elastic to brittle. The value of the glass temperature increases with the increasing number of crosslinks. Examples of elastomers are butadiene resin, styrene-butadiene resin, and polyurethane resin. Raising temperature affects an increase of elasticity, caused by reducing the stiffening effects of the crosslinks and increasing the mobility of the molecule chains. On exceeding the decomposition temperature T_d, the atom bonding within and between the molecule chains will be broken, and the material will be chemically decomposed (Bhowmick and Stephens, 2000, Klein, 2011) as shown in Figure 1.4.

1.3.1.3 Thermosets

Thermosets are plastic resins with narrow crosslinked molecule chains. Examples of thermosets are epoxy resin, phenolic resin, and polyester resin. In the state of application, thermosets are hard and brittle. Because of the strong resistance of molecular movement caused by the crosslinking, mechanical strength and elasticity are not temperature dependent, as with thermoplastics or elastomers. Thermosets cannot be melted and joining by thermal processes like ultrasonic welding or laser welding is not possible. On exceeding the decomposition temperature T_d, the material will be chemical decomposed (Frank and Biederbick, 1984). A typical structure of thermosets is represented in Figure 1.4.

1.3.1.4 Polymer Compounds

Polymer compound summarizes polymer blends, copolymers, and thermoplastic elastomers (TPEs).

1.3.1.4.1 Polymer Blends

Polymer blends are combinations of different polymers (Robeson, 2007), usually mixed in molten. After solidification, the different polymeric proportions are combined by physical but not chemical reactions (Figure 1.4). The extent to which a mixture can be achieved depends on the miscibility of the polymers among each other. Chemical, thermal or mechanical properties of polymer blends are defined by the different polymers used and their proportions within the polymer blend. Polymer blends, designed from thermoplastic materials, can be joined together by thermal processes like ultrasonic or laser welding. Examples of thermoplastic polymer blends are PC/ABS, PC/ASA, or PPE/SB (Kerin, 1998).

1.3.1.4.2 Copolymers

Copolymers are built by chemical composition at least from two different monomer units. Processes for copolymers are block polymerization, group transfer polymerization, or graft copolymerization (Frank and Biederbick, 1984; Batzer, 1985). Examples of copolymers are ABS or styrene-acrylonitrile copolymers. Besides grade of polymerization, chain-length distribution, type of end groups and chain side branches, composition, and distribution of monomer units inside the molecule chain are vital to achieving specific chemical, thermal, optical, or mechanical properties of the copolymer. Especially influential on the properties is the regularity of the chain composition, which means a statistical or more regular distribution of the different monomers within the molecule chain (Osswald and Menges, 1996).

1.3.1.4.3 Thermoplastic Elastomers

Thermoplastic elastomers are materials that combine many of the attributes and features of both vulcanized thermoset rubber and thermoplastic materials. The processing and behavior of TPE materials classify them as belonging to a group of materials between thermoplastics and elastomers. They form an independent class of materials and close the gap between stiff thermoplastics and vulcanized elastomers.

There are six generic classes of commercial TPEs included:

1. Styrenic block copolymers, TPS (TPE-s)
2. Thermoplastic polyolefin elastomers, TPO (TPE-o)
3. Thermoplastic Vulcanizates, TPV (TPE-v or TPV)
4. Thermoplastic polyurethanes, TPU (TPU)
5. Thermoplastic copolyester, TPC (TPE-E)
6. Thermoplastic polyamides, TPA (TPE-A)

Thermoplastic elastomers described as TPS are compounds based on styrene-butadiene-styrene (SBS) or styrene-ethylene-butylene-styrene (SEBS). SBS is based on two-phase block copolymers with hard and soft segments. The styrene end blocks provide the thermoplastic properties, and the butadiene mid-blocks provide the elastomeric properties. SBS is probably the highest volume TPS material produced and is commonly used in footwear, adhesives, and lower specification seals and grips, where resistance to chemicals and aging are not critical. SBS, when hydrogenated, becomes SEBS, as the elimination of the C=C bonds in the butadiene component generated ethylene and butylene mid-blocks. SEBS is characterized by improved heat resistance, mechanical properties, and chemical resistance.

TPO compounds are PP resin blends and un-crosslinked ethylene propylene diene monomer (EPDM) rubber and polyethylene. They are characterized by high impact resistance, low density, and good chemical resistance. They are used in applications requiring increased toughness and durability over the conventional PP copolymers, such as automotive bumpers and dashboards. The properties are restricted to the high end of the hardness scale, typically >80 Shore A, and with limited elastomeric properties. TPOs can be easily processed by injection molding, extrusion, or blow molding.

TPV compounds are the next step up in performance from TPO. These, too, are PP and EPDM rubber compounds; however, they have been dynamically vulcanized during the compounding step. They were originally conceived to bridge thermoplastic materials and vulcanized EPDM.

They have seen strong growth in automotive seals, pipe seals, and other applications requiring a heat resistance of up to 120°C. Shore hardness values range typically from 45A to 45D. TPVs also lend themselves to under-bonnet automotive applications where improved temperature and oil resistance are required.

Thermoplastic polyurethane is a plastic category created when a polyaddition reaction occurs between di-isocyanate and one or more diols. They can be used as a soft engineering plastic or a hard rubber replacement.

Copolyester elastomers are high-performance, high-temperature elastomers with many thermoset rubber features but the processing ease of engineering plastics. Typical applications include automotive, appliance, electrical, consumer, furniture, and lawn and garden components.

Thermoplastic polyamide elastomers (PAE, TPA, TPE-A, PEBA, COPA) are high-performance TPEs block copolymers based on nylon and polyethers or polyesters. They are used mainly in areas where other thermoplastic elastomers cannot compete or perform, especially at a lower temperature.[9, 10, 11]

1.3.1.4.4 Polymer or Plastic Composites

A polymer or plastic composite is a multi-phase material in which reinforcing fillers are integrated with a polymer matrix (thermoplastic or thermosets), resulting in synergistic mechanical properties that cannot be achieved from either component alone. Examples include fiber-reinforced plastics, sheet molding compounds, bulk molding compounds,

---

[9] https://www.entecpolymers.com/products/resin-types/thermoplastic-copolyester-elastomer-tpc-et.
[10] https://www.iso.org/obp/ui/#iso:std:iso:18064:ed-2:v1:en.
[11] https://www.hexpol.com/tpe/resources/tpe-academy/what-is-tpe/.

pre-preg materials, glass mat plastics, weave-reinforced glass mat plastics, low-density composites, and composite sandwich panels glass mat plastics. Importantly, polymer compounds with thermoplastic matrices usually can be melted by thermal processes like welding, but not polymer compounds with thermoset matrix.

Some of the most common matrices are polyacetal, polyetheretherketone, epoxy, fluoropolymers, and phenolics.

Polyoxymethylene or Polyacetal is a thermoplastic used in precision parts that require high stiffness, low friction, and excellent dimensional stability. It provides a higher strength material than polyethylene-type polymers; however, Polyacetal materials are susceptible to oxidation at elevated temperatures.

Polyether ether ketone is a colorless organic polymer thermoplastic. It has excellent mechanical and chemical resistance properties that are retained at high temperatures. It is highly resistant to thermal degradation as well. It is used extensively in the aerospace, automotive, electronic, and chemical process industries.

Epoxy resins that belong to the thermosetting class exhibit high strength and low shrinkage during curing. They are known for their toughness and resistance to chemical and environmental damage. Primarily they are used for coating purposes but depending on the formulation, epoxy resins are used as potting agents, resin binders, or laminating resins in fiberglass or composite construction. They are also used as encapsulates, electrical conductors in microelectronic packaging, and adhesives in structural bonding applications.

Fluoropolymers can be mechanically characterized as thermosets or thermoplastics. Commonly fluoropolymers including polytetrafluoroethylene and polyvinylidene fluoride are used in applications requiring superior chemical resistance or low friction.

Phenolics are thermosetting molding compounds and adhesives that offer strong bonds and provide good resistance to high temperatures. Phenolic resin adhesives made from chemicals of the phenol group and formaldehyde are generally the most durable. Phenolic resins are available in liquid, powder, and film forms. Special phenolic resins are available that harden at moderate temperatures when mixed with suitable accelerators. Urea-formaldehyde resins can harden rapidly at moderate temperatures but generally do not have the properties of phenolic resins. Melamine resins have excellent dielectric properties.

Most matrices can be strengthened with fibers, fillers, particulates, powders, and other matrix reinforcements to improve strength and/or stiffness. Fibers are usually chopped, wound, or woven and made of materials such as fabric, metal, glass, or fiberglass. Particulates vary in terms of shape and size. Powders are usually made of carbon, graphite, silicates, ceramics, and other organic or inorganic materials. Some matrix reinforcements provide improved electrical conductivity while others offer improved thermal conductivity.

1.3.2 CLASSIFICATION BASED ON APPLICATION OF MATERIAL

Based on the application of material, plastics (most probable thermoplastics) are classified into four types included standard or commodity plastics, engineering plastics, special plastics, and high-performance plastics.

1.3.2.1 Standard or Commodity Plastics

In applications where mechanical properties and operating environment are not critical, plastics with wide-ranging applications, termed commodity plastics, are considered the most significant commercially. Consequently, these common plastics have
an exceptionally high production rate. As a result of their worldwide abundance,
commodity plastics are the most widespread plastics found littering the aquatic
environment. Individual commodity plastics have been allocated specific designator codes by the international standards organization ASTM (American Society for
Testing and Materials) as indicated in Figure 1.6. These codes enable these plastic
materials to be easily identified and separated (Harrison, 2014).

1.3.2.1.1   Class-1, Polyethylene Terephthalate

Polyethylene terephthalate (PET) is a chemically stable polyester and its use has risen
dramatically in the last few decades with a multitude of applications, ranging from
food and drink containers to the manufacture of electronic components and as fibers
in clothes. Often, recycled PET bottles are used to make fleece garments, as well as
plastic bottles. Indeed, one of the most common uses of PET is in the manufacture of
drinking water bottles (Harrison, 2014; ICES, 2015).

1.3.2.1.2   Class-2, High-Density Polyethylene

HDPE has a closer-packed structure, making them dense and, thus, stronger and
thicker than PET. The densities of HDPE and LDPE plastics can vary by 0.01–0.05 g
mL^{-1} values, but both are less dense than water. HDPE is commonly used in a grocery bag, opaque milk, juice container, shampoo bottles, and medicine bottle.

   Not only recyclable, but HDPE is also relatively more stable than PET. It is considered a safer option for food and drink; however, some studies have shown that it can
leach estrogen-mimicking additive chemicals that could disrupt the human's hormonal
system when exposed to ultraviolet light (ICES, 2015; Harris and Walker, 2010).

1.3.2.1.3   Class-3, Polyvinyl Chloride

Pure PVC is a brittle white-colored substance without additives. During manufacture, PVC is produced in two varieties; rigid and flexible. The rigid type is a durable

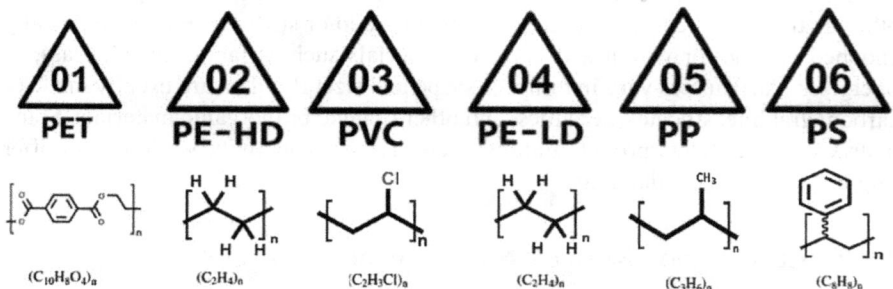

01	02	03	04	05	06
PET	PE-HD	PVC	PE-LD	PP	PS
$(C_{10}H_8O_4)_a$	$(C_2H_4)_n$	$(C_2H_3Cl)_a$	$(C_2H_4)_n$	$(C_3H_6)_a$	$(C_8H_8)_n$

FIGURE 1.6 Commodity plastics with specific designator codes, chemical structure, and
chemical formula. (Derived from Harrison, 2014; Crawford and Quinn, 2017; ICES, 2015.)

hard material that exhibits fire-retardant properties and demonstrates resistance to chemical degradation and weathering. However, it tends to have low impact resistance at room temperature (21°C). Any weathering to the material tends to only occur on the material's surface. Phthalate plasticizers create the plasticized flexible form of PVC during manufacture to soften the polymer. When PVC is compounded with wood flour fillers, the resulting formulation may be susceptible to biological attack. Consequently, antimicrobial additives may be added were necessary to combat bio-degradative effects.

The presence of chlorine provides fire retarding properties and imparts an ignition temperature of up to 455°C, as well as protecting against oxidation (Tongesayi and Tongesayi, 2015; Harris and Walker, 2010; Davis et al., 2010).

1.3.2.1.4   Class-4, Low-Density Polyethylene

Low-density polyethylene (LDPE) has a branched-chain structure making it less dense and less crystalline (structurally ordered) and thus a generally thinner, more flexible form of polyethylene. LDPE is mostly used for bags (grocery, dry cleaning, bread, frozen food bags, newspapers, garbage), plastic wraps; coatings for paper milk cartons, hot & cold beverage cups; some squeezable bottles (honey, mustard), food storage containers, container lids. Also used for wire and cable covering. LDPE is considered a safer plastic option for food and drink use but is quite difficult to be recycled[12] (Harris and Walker, 2010).

1.3.2.1.5   Class-5, Polypropylene

Polypropylene is regarded as one of the lightest and most versatile polymers resistant to many acids, alkalis and have low density, high stiffness, good balance of impact strength versus rigidity, heat resistance and exhibits good transparency. It can undergo many manufacturing processes, such as injection molding, general-purpose extrusion, extrusion blow molding, and even expansion molding. Although existing in three different tactic forms, commercial polypropylene is typically a mixture of 75% isotactic and 25% atactic. Demand for polypropylene is rapidly on the increase and consequently, it is one of the most common types of microplastic found in the marine environment[13] (Rochman et al., 2014).

1.3.2.1.6   Class-6, Polystyrene

Rigid varieties of this class are transparent, hard, and brittle. Unfilled types have a sparkling crystal-like appearance. High impact polystyrene is produced by blending with a butadiene copolymer or rubber, increasing impact resistance and toughness. Upon heating, polystyrene tends to depolymerize to styrene. A more rigid form of polystyrene can be produced by adding *p*-divinylbenzene during manufacture. It results in crosslinking and has the copolymer poly(styrene-co-divinylbenzene), which exhibits increased resistance to organic solvents. This crosslinked polystyrene is typically used to produce ion exchange resins in the form of spherical beads 200–1,200 µm in size. Polystyrene can also be mixed with a volatile solvent, typically

---

[12] https://waste4change.com/blog/7-types-plastic-need-know.
[13] https://waste4change.com/blog/7-types-plastic-need-know.

5% pentane by weight. When this mixture is heated, the pentane expands and bubbles to produce low-density polystyrene foam termed Styrofoam. Alternatively, small spherical beads of polystyrene can be expanded with pentane to around 40 times their original size to have a material composed of low-density spherical beads of polystyrene 0.5–1.0 mm in size, termed styropor. Foamed or expanded, varieties of polystyrene are often found floating on the surface of aquatic environments, while solid polystyrene, being slightly denser than that of water, is generally found below the surface[14] (Addamo et al., 2017).

1.3.2.1.7 Class-7, Other Plastics

The most prominent are polycarbonate, poly(methyl acrylate), poly(methyl methacrylate), polytetrafluoroethylene, ABS, polyamide, polyhydroxybutyrate, polycaprolactone, polylactic acid, etc. Polycarbonate is a thermoplastic that contains organic functional groups connected by carbonate groups. Due to the long molecular chain, the material can be easily thermoformed. Often, polycarbonate is blended with other plastics, such as ABS and rubber.[15]

Poly(methyl acrylate) (**PMA**) is a hydrophobic synthetic acrylate polymer. PMA, though softer than PMMA, is tough, leathery, and flexible. It has a low glass-transition temperature of about 10°C (12.5°C in the case of PMA). High-energy radiation leads to cross-linking in PMA. However, in PMMA, a compound similar to PMA, degradation occurs instead. It is soluble in dimethyl sulfoxide. PMA is water-sensitive and is not stable against alkalies. It is used as a macroinitiator to initiate the polymerization of HEMA and DMAEMA. Also used in leather finishing and textiles (Ciullo, 1996; Brydson, 1999; Guice, 2008).

Poly(methyl methacrylate) (**PMMA**), also known as acrylic or acrylic glass, is a transparent and rigid thermoplastic material widely used as a shatterproof replacement for glass. PMMA has many technical advantages over other transparent polymers (PC, polystyrene, etc.); a few of them include high resistance to UV light and weathering and excellent light transmission. PMMA or poly (methyl 2-methyl propionate) is produced from monomer methyl methacrylate. PMMA is a transparent, colorless polymer available in pellet, small granules, and sheet forms, formed with all thermoplastic methods (including injection molding, compression molding, and extrusion). The highest quality PMMA sheets are produced by cell casting, but in this case, the polymerization and molding steps occur concurrently. It is commonly called acrylic glass. The strength of the material is higher than molding grades owing to its extremely high molecular mass. Rubber toughening has been used to increase the toughness of PMMA owing to its brittle behavior in response to applied loads.[16]

Polytetrafluoroethylene (**PTFE**) is a highly crystalline robust, unique plastic with a high melting point. The material has a propensity to exhibit considerable elastic deformation under load. Consequently, in applications that require a frictionless surface and resistance to load, such as bearing surfaces, the deformation characteristics of PTFE are improved with the use of additives and fillers. PTFE has a high

---

[14] https://polymerdatabase.com/polymer%20classes/Polystyrene%20type.html.
[15] http://wwwcourses.sens.buffalo.edu/ce435/PC_CB.pdf.
[16] https://omnexus.specialchem.com/selection-guide/polymethyl-methacrylate-pmma-acrylic-plastic.

molecular weight and is generally an unreactive substance due to the highly stable structure's fluorine and carbon bonds. Furthermore, the high electronegativity of the fluorine atoms gives the plastic an extremely water repellent surface and outstanding non-stick properties (Radulovic and Wojcinski, 2014).

ABS is a high luster opaque substance with high surface quality. The material's high strength, hardness, impact resistance, and rigidity are attributed to the copolymerization of styrene and acrylonitrile, while the toughness is attributed to fine particles of polybutadiene rubber uniformly interspersed throughout the styrene-acrylonitrile copolymer matrix. The ratios of styrene, acrylonitrile, and polybutadiene can be altered to produce different grades of ABS with different properties. Furthermore, ABS is often used in blends with other materials to impart toughness or with PVC to gain flame resistance. Although ABS is an engineering plastic and is a standard plastic used in many electrical appliance cases, the substance is prone to fire and smoking. Consequently, ABS requires the addition of flame retardants to suppress this weakness or blend with PVC. In the environment, ABS is highly susceptible to weathering and has a density greater than that of seawater, tending to exist below the surface in aquatic environments (Davis et al., 2010; Yang et al., 2014; Harris and Walker, 2010; Collard et al., 2015; Davis and Sims, 1983).

Nylons (polyamides) are one of the most widely used materials worldwide and are generally known as Nylon 6 or Nylon 6,6. Synthesis of Nylons is achieved by a polymerizing condensation reaction in which an amino group on one monomer reacts with the carboxyl group on another to create an amide bond and produce a polymeric translucent fibrous-like substance. In Nylon 6,6, the name derives from the fact that the two monomers of Nylon 6,6 (hexamethylenediamine and adipic acid) each contain six carbon atoms. Nylon fibers exhibit elasticity and have exceptional strength and toughness, and are more robust than fibers of PET. Furthermore, Nylons have excellent wear resistance (Andrady, 2011; Hauser and Calafat, 2005).

Polyhydroxybutyrate (PHB) is the most popular biodegradable thermoplastic, belonging to the polyester class of compounds. PHB is insoluble in water and thus has a better resistance against hydrolytic degradation than other biodegradable plastics, which are soluble in water or sensitive to moisture. Furthermore, the material exhibits good resistance to ultraviolet light. Many items are available that are made from PHB, such as shampoo bottles, cups, and golf tees. While in Japan, there is an extensive range of PHB disposable razors on the market. PHB is also used within the human body as sutures that break down naturally due to its non-toxic degradative properties (Chamas et al., 2020).

Polycaprolactone is a biodegradable plastic with a low melting point of 60°C and is therefore unsuitable for high-temperature applications. However, it is occasionally blended with other plastics to improve impact resistance or plasticize PVC. Due to its biodegradative properties in the human body, which occurs slower than polylactic acid (PLA), there is considerable research underway to develop implantable devices or sutures that can remain in the body for a long period before breaking down when no longer required. Furthermore, the encapsulation of drugs with polycaprolactone for controlled and targeted drug delivery systems has been successfully accomplished (Liu et al., 2019b).

PLA is the second most widely used biodegradable plastic. It is derived from lactic acid, which can be produced by fermentation of renewable agricultural produce, such as sugar cane or corn starch. PLA has many applications, such as product packaging material, tableware, and feedstock material for desktop 3D printers. Furthermore, owing to the substance's ability to degrade to the innocuous lactic acid monomer, PLA is used within the human body for medical implants, such as pins, rods, and screws. In the body, PLA completely degrades within 6–24 months, depending on the precise composition. The slow rate of degradation in the body is advantageous in weight-bearing structures like bone since loading is gradually returned to the body as the polymer degrades and the body heals. Furthermore, lactic acid can be copolymerized with glycolic acid to create poly(lactic-co-glycolic acid) (PLGA), a biodegradable and biocompatible polymer. PLGA can be used for the targeted delivery of drugs within the body, such as the antibiotic amoxicillin, by creating microplastic-sized (~60 μm) solid PLGA microcapsules, which are loaded with the active drug are produced using micro-jetting technology. Once inside the body, the solid PLGA microcapsules degrade when exposed to water via hydrolysis of their ester linkages, releasing the active drug. When these biodegradable plastic materials are discarded into the environment, complete biodegradation takes about 2 weeks in the case of discard into a sewage treatment facility and approximately 2 months in the case of discard into the soil or aquatic environments (Sheikh et al., 2015; Talsness et al., 2009). Figure 1.7 indicates class 7 of commodity plastics with specific designator codes, chemical structure, and chemical formula.

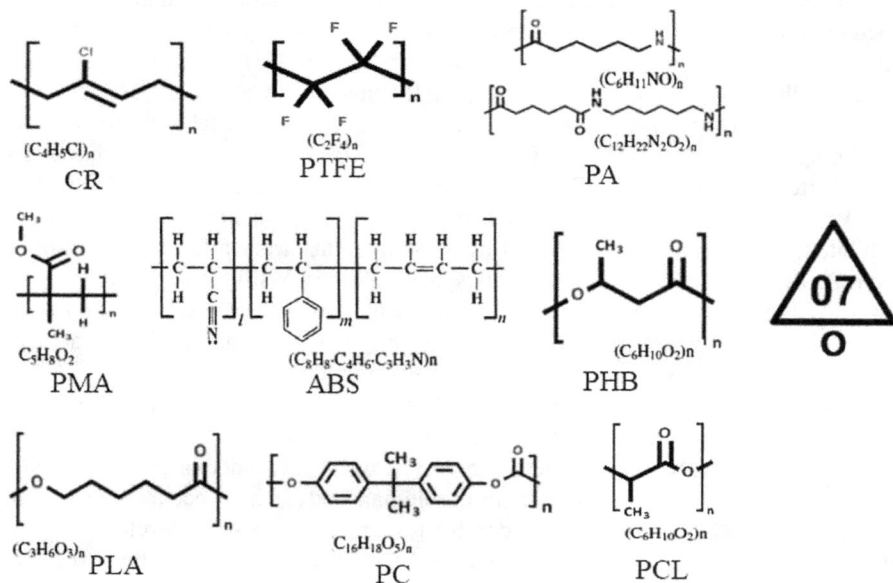

FIGURE 1.7 Class 7 of commodity plastics with specific designator codes, chemical structure, and chemical formula.

1.3.2.2 Engineering Plastics

Engineering plastics are used when good structural, transparency, self-lubrication, and thermal properties are needed. Some examples are PA, polyacetal (POM), PC, PET, polyphenylene ether, and polybutylene terephthalate.[17]

1.3.2.3 Special Plastics

They have a specific property to an extraordinary degree, such as PMMA, which has high transparency and light stability, or polytetrafluoroethylene (Teflon), which has good resistance to temperature and chemical products.

1.3.2.4 High-Performance Plastics

Mostly thermoplastic with high heat resistance. In other words, they have good mechanical resistance to high temperatures, particularly up to 150°C. PI, polysulfone, polyethersulfone, polyarylsulfone, polyphenylene sulfide, and liquid crystal polymers are high-performance plastics.[18]

1.4 ENVIRONMENTAL IMPACT AND ANALYTICAL CHALLENGES; AN OVERVIEW

Determining the fate of (micro/nano)plastics in the environment is inherently difficult. This is mostly due to the multiplicity of sources and routes of entry into the environment and the timescales necessary to determine their degradation pathways. For smaller particles, this is due to their size as well. Generally, in response to the issue of microplastic contamination of aquatic environments, policy advisors and resource managers are asking variations of the following questions: (i) what and how much microplastics are in our waters? (ii) Where are the microplastics coming from? (iii) What harm do the microplastics cause? (iv) What can be done to reduce the presence of microplastics in the environment? Studies of microplastics to date, and those underway, aim to answer these questions and to help guide policy development and implementation of resource management activities to address the issue (Helm, 2017). Microplastics have been identified across the globe, from the Arctic to the Antarctic and from the surface to the depths (benthos). But microplastics are also found in rivers and lakes, in agricultural soils, sediments, and even in the atmosphere, both in indoor and outdoor environments (Woodall et al., 2014; Chouchene et al., 2019; Cincinelli et al., 2017; Zhang et al., 2016; Vianello et al., 2019; Boots et al., 2019). Once in the environment, plastics can undergo degradation through abiotic and/or biotic processes, and the former is an essential first step that precedes the latter. In other words, biodegradation mechanisms require an initial abiotic degradation process. This yields materials of diminished structural and mechanical integrities, resulting in particles with higher surface-area-to-volume ratios, amenable to microbial action (Alshehrei, 2017) (see Chapters 1, 2 and 3).

After biotic or abiotic degradation, the interaction of organisms with plastic debris results in a wide range of consequences, both direct and indirect, including the potential occurrence of sub-lethal effects, which, owing to their uncertainty, may be of considerable concern. Broadly, the presence of larger plastic materials in the ocean

---

[17] https://www.aimplas.net/blog/plastics-identification-and-classification/.
[18] https://www.aimplas.net/blog/plastics-identification-and-classification/.

may result in entanglement and ingestion, the potential creation of new habitats, and dispersal via rafting, including transport of invasive species. Entanglement and ingestion frequently cause harm or death, although gathered data appears to suggest that entanglement is far more fatal (79% of all cases) than ingestion (4% of all cases) (Gall and Thompson, 2015). Debris may also constitute new habitats and derelict fishing gear, causing death by "ghost fishing" and constituting new habitats for invertebrates (Good et al., 2010). The dispersal of species in the marine environment, particularly species with no pelagic larval stage, has increased in recent decades. Highly dependent on oceanic currents, numerous species have always rafted on natural materials such as wood, but industrialization and the continuous increase of plastic debris in the oceans suggest that rafting is playing an active role in their scattering (Gall and Thompson, 2015). This holds true for invasive species as well. A clear example is the presence of a ciliate, Halofolliculina, a pathogen that may be the culprit of the eroding skeletal disease that has affected Caribbean and Hawaiian corals (Goldstein et al., 2014).

Due to their small size, microplastics may be ingested by multiple organisms, such as planktonic and higher organisms, including mammals, birds, and fish. Although the exact mechanisms of toxicity of these materials are still ill-understood, the effects are potentially due to either (i) ingestion induced stress, such as physical blockage, energy expenditure for egestion, and false satiety; (ii) leakage of chemicals, such as additives, from plastics; and (iii) exposure to contaminants adsorbed (and subsequently released) by microplastics such as persistent organic pollutants (POPs) (da Costa, 2017). Cnidarians, annelids, ciliates, rotifers, copepods, amphipods, euphausiids, mussels, barnacles, tunicates, birds, and fish have all been demonstrated to ingest these small-sized polymers within laboratory settings (Browne et al., 2008; Duis and Coors, 2016; da Costa, 2019) (see Chapter 4). As such, the quantification of these materials is rather difficult, particularly given that, especially for smaller sized plastics, there is a lack of standardized methods for their sampling, unit normalization, data expression, and quantification, as well as identification (see Chapter 5). Methods of analysis need to ensure that they provide the appropriate information, rigor, and accuracy with an efficient understanding of fundamental principles of analysis (Helm, 2017). Many of the sampling, extraction, and analysis method improvements and considerations discussed within this book will contribute to more efficient and effective monitoring and assessment of microplastics within the aquatic environment (see Chapters 6–8).

ACKNOWLEDGMENT

This work was supported by a grant from the National Research Foundation of Korea (NRF) grant funded by the Korea government (MSIT) (No. NRF-2020R1A2C1013851).

REFERENCES

Addamo, A. M., Laroche, P., Hanke, G., 2017. Top marine beach litter items in Europe: A review and synthesis based on beach litter data. *JRC Technical Reports*, EUR 29249 EN. Publications Office of the European Union, Luxembourg. publications.jrc. ec.europa.eu/repository/bitstream/JRC108181/technical_report_top_marine_litter_items_eur_29249_en_pdf.pdf.

Alshehrei, F., 2017. Biodegradation of synthetic and natural plastic by microorganisms. *Journal of Applied & Environmental Microbiology*, 5(1), 8–19.

Andrady, A. L., 2011. Microplastics in the marine environment. *Marine Pollution Bulletin*, 62(8), 1596–1605.

Andrady, A. L., 2015. *Plastics and Environmental Sustainability*. John Wiley & Sons. ISBN: 978-1-118-31260-5.

Avramova, N., Fakirov, S., 1981. Melting behaviour of drawn and undrawn annealed nylon6. *Acta Polymerica*, 32(6), 318–322.

Batzer, H., 1985. Polymer materials. *Chemistry and Physics*, 1, 157. Thieme Medical Press, New York.

Baysal, B., Tobolsky, A. V., 1952. Rates of initiation in vinyl polymerization. *Journal of Polymer Science*, 8(5), 529–541.

Billmeyer, F. W., 1984. *Textbook of Polymer Science*. John Wiley & Sons, Hoboken. ISBN: 0471031968.

Boots, B., Russell, C. W., Green, D. S., 2019. Effects of microplastics in soil ecosystems: Above and below ground. *Environmental Science & Technology*, 53(19), 11496–11506.

Braun, D., 1985. Plastics: Plastics compendium. In: A. Franck, K.-H. Biederbick. Vogel-Buchverlag Würzburg 1984, (Eds.). pp. 346. ISBN: 3-8023-0135-8.

Bhowmick, A.K., Stephens, H.L. 2000. *Handbook of Elastomers*, 2nd edn, CRC-Press, Boca Raton.

Browne, M. A., Dissanayake, A., Galloway, T. S., Lowe, D. M., Thompson, R. C., 2008. Ingested microscopic plastic translocates to the circulatory system of the mussel, Mytilus edulis (L.). *Environmental Science & Technology*, 42(13), 5026–5031.

Brydson, J. A., 1989. *Plastic Materials*, 5th edn. Butterworths, London.

Brydson, J. A., 1999. *Plastics Materials*. Butterworth-Heinemann. p. 423. ISBN: 978-0-7506-4132-6.

Chamas, A., Moon, H., Zheng, J., Qiu, Y., Tabassum, T., Jang, J. H., Abu-Omar, M., Scott, S. L., Suh, S., 2020. Degradation rates of plastics in the environment. *ACS Sustainable Chemistry & Engineering*, 8(9), 3494–3511.

Chandler, A. D., 2005. *Shaping the Industrial Century. The Remarkable Story of the Evolution of the Modern Chemical and Pharmaceutical Industries*. Harvard University Press, Cambridge, MA.

Chouchene, K., da Costa, J. P., Wali, A., Girão, A. V., Hentati, O., Duarte, A. C., Ksibi, M., 2019. Microplastic pollution in the sediments of Sidi Mansour Harbor in Southeast Tunisia. *Marine Pollution Bulletin*, 146, 92–99.

Cincinelli, A., Scopetani, C., Chelazzi, D., Lombardini, E., Martellini, T., Katsoyiannis, A., Corsolini, S., 2017. Microplastic in the surface waters of the Ross Sea (Antarctica): Occurrence, distribution and characterization by FTIR. *Chemosphere*, 175, 391–400.

Ciullo, P. A., 1996. *Industrial Minerals and Their Uses: A Handbook and Formulary*. Elsevier. p. 115. ISBN: 978-0-8155-1408-4.

Collard, F., Gilbert, B., Eppe, G., Parmentier, E., Das, K. 2015. Detection of anthropogenic particles in fish stomachs: An isolation method adapted to identification by Raman spectroscopy. *Archives of Environmental Contamination and Toxicology*, 69(3), 331–339.

Cossee, P., 1964. Ziegler-Natta catalysis I. Mechanism of polymerization of α-olefins with Ziegler-Natta catalysts. *Journal of Catalysis*, 3(1), 80–88.

Crawford, C. B., Quinn, B., 2017. *Microplastic Pollutants*. Elsevier Science. ISBN: 9780128094068.

da Costa, J. P., Santos, P. S., Duarte, A. C., Rocha-Santos, T., 2016. (Nano) plastics in the environment–sources, fates and effects. *Science of the Total Environment*, 566, 15–26.

da Costa, J. P., 2017. Microplastics–occurrence, fate and behaviour in the environment. In: A.P.T. Rocha-Santos, A.C. Duarte (Eds.), *Comprehensive Analytical Chemistry*. Elsevier, Amsterdam. pp. 1–24.

da Costa, J. P., Nunes, A. R., Santos, P. S., Girao, A. V., Duarte, A. C., Rocha-Santos, T., 2018. Degradation of polyethylene microplastics in seawater: Insights into the environmental degradation of polymers. *Journal of Environmental Science and Health, Part A*, 53(9), 866–875.

da Costa, J. P., 2019. Nanoplastics in the environment. In: Harrison and Hester (Eds.), *Plastics and the Environment*. The Royal Society of Chemistry, Burlington House, Piccadilly, London. pp. 82–105.

Davis, A., Sims, D., 1983. *Weathering of Polymers*. Applied Science Publishers Ltd. ISBN: 0-85334-226-1.

Davis, M. E., Zuckerman, J. E., Choi, C. H. J., Seligson, D., Tolcher, A., Alabi, C. A., Yen, Y., Heidel, J. D., Ribas, A., 2010. Evidence of RNAi in humans from systemically administered siRNA via targeted nanoparticles. *Nature*, 464(7291), 1067–1070.

Duis, K., Coors, A., 2016. Microplastics in the aquatic and terrestrial environment: Sources (with a specific focus on personal care products), fate and effects. *Environmental Sciences Europe*, 28(1), 1–25.

Edmondson, S., Gilbert, M., 2017. The chemical nature of plastics polymerization. In *Brydson's Plastics Materials*, 8th edn. Elsevier Science. ISBN: 978-0-323-35824-8.

Elias, H. G., 1993. *An Introduction to Plastics*. VCH, Weinheim. ISBN: 3-527-28578-4.

Flory, P. J., 1953. *Principles of Polymer Chemistry*. Cornell University Press, Ithaca and London, New York. ISBN: 0-8014-0134-8.

Frank, A. and Biederbick, K. 1984. *Kunststoffkompendium*. Vogel-Buchverlag, Wurzburg.

Fried, J. R., 1995. *Polymer Science and Technology*. Prentice Hall, Upper Saddle River, NJ.

Gall, S. C., Thompson, R. C., 2015. The impact of debris on marine life. *Marine Pollution Bulletin*, 92(1–2), 170–179.

Gigault, J., Ter Halle, A., Baudrimont, M., Pascal, P. Y., Gauffre, F., Phi, T. L., Reynaud, S., 2018. Current opinion: What is a nanoplastic? *Environmental Pollution*, 235, 1030–1034.

Goldstein, M. C., Carson, H. S., Eriksen, M., 2014. Relationship of diversity and habitat area in North Pacific plastic-associated rafting communities. *Marine Biology*, 161(6), 1441–1453.

Good, T. P., June, J. A., Etnier, M. A., Broadhurst, G., 2010. Derelict fishing nets in Puget Sound and the Northwest Straits: Patterns and threats to marine fauna. *Marine Pollution Bulletin*, 60(1), 39–50.

Goodman, I., Rhys, J. A., 1965. *Polyesters, Vol. 1: Saturated Polymers*. Iliffe Books, London.

Gregg, R. A., Mayo, F. R., 1947. Chain transfer in the polymerisation of styrene iii. The reactivities of hydrocarbons toward the styrene radical. *Discussions of the Faraday Society*, 2, 328–337.

Guice, K. B., 2008. *Synthesis & Characterization of Temperature- and PH-responsive Nanostructures Derived from Block Copolymers Containing Statistical Copolymers of HEMA and DMAEMA*. p. 29. ISBN: 978-0-549-63651-9.

Harris, F. W., Norris, S. O., Lanier, L. H., Reinhardt, B. A., Case, R. D., Varaprath, S. Padaki, S. M., Torres, M., Feld, W. A., 1982. In: K. L. Mittal (Ed.), *Polyimides*, Vol. 1. pp. 1–5. Plenum, New York.

Harris, M. E., Walker, B., 2010. A novel, simplified scheme for plastics identification: JCE classroom activity 104. *Journal of Chemical Education*, 87(2), 147–149.

Harrison, R. M., 2014. *Pollution: Causes, Effect and Control*, 5th edn. Royal Society of Chemistry Publishing, Cambridge, UK. ISBN: 978-1-84973-648-0.

Hauser, R., Calafat, A. M., 2005. Phthalates and human health. *Occupational and Environmental Medicine*, 62(11), 806–818.

Helm, P. A., 2017. Improving microplastics source apportionment: A role for microplastic morphology and taxonomy? *Journal of Analytical Methods in Chemistry*, 9, 1328.

ICES. 2015. OSPAR request on development of a common monitoring protocol for plastic particles in fish stomachs and selected shellfish on the basis of existing fish disease surveys. ICES Advice. https://www.ices.dk/sites/pub/Publication%20Reports/Advice/2015/Special_Requests/OSPAR_PLAST_advice.pdf.

Kataoka, T., Hinata, H., Kato, S., 2013. Analysis of a beach as a time-invariant linear input/output system of marine litter. *Marine Pollution Bulletin*, 77(1–2), 266–273.

Kerin, A.J., Wisnom, M.R., Adams, M.A. 1998. The compressive strength of articular cartilage. *Proceedings of the Institution of Mechanical Engineers*, 212, 273–280.

Klein, I. R., 2011. *Laser Welding of Plastics: Materials, Processes and Industrial Applications*. John Wiley & Sons, Hoboken, NJ. ISBN: 9783527409723.

Krentsel, B. A., Kissin, Y. V., Kleiner, V. J., Strotskava, L., 1997. *Polymers and Co-polymers of α-Olefins*. Hanser, Munich.

Kresser, T. O. J., 1961. *Polyethylene*. Reinhold, New York.

Liu, K., Wu, T., Wang, X., Song, Z., Zong, C., Wei, N., Li, D., 2019a. Consistent transport of terrestrial microplastics to the ocean through atmosphere. *Environmental Science & Technology*, 53(18), 10612–10619.

Liu, S., Wu, G., Chen, X., Zhang, X., Yu, J., Liu, M., Zhang, Y., Wang, P., 2019b. Degradation behavior in vitro of carbon nanotubes (CNTs)/poly(lactic acid) (PLA) composite suture. *Polymers*, 11, 1015. doi: 10.3390/polym11061015.

Matyjaszewski, K., Davis, T. P., 2003. *Handbook of Radical Polymerization*. John Wiley & Sons, Hoboken. ISBN: 0471461571.

Millet, H., Vangheluwe, P., Block, C., Sevenster, A., Garcia, L., Antonopoulos, R., 2018. The nature of plastics and their societal usage. In: *Plastics and the Environment*. pp. 1–20. doi:10.1039/9781788013314-00001. eISBN: 978-1-78801-331-4.

Ng, K. L., Obbard, J. P., 2006. Prevalence of microplastics in Singapore's coastal marine environment. *Marine Pollution Bulletin* 52(7), 761–767.

Osswald, T.A., Menges, G. 1996. *Material Science of Polymers for Engineers*, Hanser Publishers, New York.

Radulovic, L. L., Wojcinski, Z. W., 2014. *Encyclopedia of Toxicology*, 3rd edn. Elsevier Science. ISBN: 978-0-12-386455-0.

Robeson, L.M. 2007. *Polymer Blends: A Comprehensive Review*. Hanser Verlag Munich.

Rochman, C. M., Manzano, C., Hentschel, B. T., Simonich, S. L. M., 2013. Polystyrene plastic: A source and sink for polycyclic aromatic hydrocarbons in the marine environment. *Environmental Science & Technology*, 47(24), 13976–13984.

Rochman, C. M., Hentschel, B. T., The, S. J., 2014. Long-term sorption of metals is similar among plastic types: Implications for plastic debris in aquatic environments. *PLoS One*, 9(1), e85433.

Sheikh, Z., Najeeb, S., Khurshid, Z., Verma, V., Rashid, H., Glogauer, M., 2015. Biodegradable materials for bone repair and tissue engineering applications. *Materials (Basel)*, 8(9), 5744–5794.

Skrovanek, D. J., Painter, P. C., Coleman, M. M., 1986. Hydrogen bonding in polymers. 2. Infrared temperature studies of nylon 11. *Macromolecules*, 19(3), 699–705.

Sommer, F., Dietze, V., Baum, A., Sauer, J., Gilge, S., Maschowski, C., Gieré, R., 2018. Tire abrasion as a major source of microplastics in the environment. *Aerosol and Air Quality Research*, 18(8), 2014–2028.

Stephens, B., Azimi, P., El Orch, Z., Ramos, T., 2013. Ultrafine particle emissions from desktop 3D printers. *Atmospheric Environment*, 79, 334–339.

Talsness, C. E., Andrade, A. J., Kuriyama, S. N., Taylor, J. A., vom Saal, F. S., 2009. Components of plastic: Experimental studies in animals and relevance for human health. *Philosophical Transactions of the Royal Society B-Biological Sciences*, 364(1526), 2079–2096.

Tongesayi, T., Tongesayi, S., 2015. *Contaminated Irrigation Water and the Associated Public Health Risks. Food, Energy, and Water*. Elsevier, Amsterdam. pp. 349–381.

Tsarevsky, N. V., Sumerlin, B. S., 2013. *Fundamentals of Controlled/Living Radical Polymerization*. Royal Society of Chemistry, Cambridge, UK. ISBN: 1849734259.

Vert, M., Doi, Y., Hellwich, K. H., Hess, M., Hodge, P., Kubisa, P., Schué, F., 2012. Terminology for biorelated polymers and applications (IUPAC Recommendations 2012). *Pure and Applied Chemistry*, 84(2), 377–410.

Vianello, A., Jensen, R. L., Liu, L., Vollertsen, J., 2019. Simulating human exposure to indoor airborne microplastics using a Breathing Thermal Manikin. *Scientific Reports*, 9(1), 1–11.

Woodall, L. C., Sanchez-Vidal, A., Canals, M., Paterson, G. L., Coppock, R., Sleight, V., Thompson, R. C., 2014. The deep sea is a major sink for microplastic debris. *Royal Society Open Science*, 1(4), 140317.

Yang, J., Yang, Y., Wu, W., Zhao, J., Jiang, L., 2014. Evidence of polyethylene biodegradation by bacterial strains from the guts of plastic-eating waxworms. *Environmental Science & Technology*, 48(23), 13776–13784.

Young, R. J., Lovell, P. A., 2011. *Introduction to Polymers*. CRC Press, Boca Raton, FL. ISBN: 13: 987-1-4398-9415-6.

Zhan, F., Zhang, H., Wang, J., Xu, J., Yuan, H., Gao, Y., Chen, J., 2017. Release and gas-particle partitioning behaviors of short-chain chlorinated paraffins (SCCPs) during the thermal treatment of polyvinyl chloride flooring. *Environmental Science & Technology*, 51(16), 9005–9012.

Zhang, H., Kuo, Y. Y., Gerecke, A. C., Wang, J., 2012. Co-release of hexabromocyclododecane (HBCD) and nano-and microparticles from thermal cutting of polystyrene foams. *Environmental Science & Technology*, 46(20), 10990–10996.

Zhang, K., Su, J., Xiong, X., Wu, X., Wu, C., Liu, J., 2016. Microplastic pollution of lakeshore sediments from remote lakes in Tibet plateau, China. *Environmental Pollution*, 219, 450–455.

2 Laws, Regulations or Policy Tools to Govern Macroplastics, Mesoplastics, Microplastics and Nanoplastics

Hyunjung Kim and Sadia Ilyas
Hanyang University

CONTENTS

DOI: 10.1201/9781003200628-2

Presently, it is estimated that plastic waste constitutes approximately 10% of the total municipal waste worldwide and that 80% of all plastic found in the world's oceans originates from land-based sources (Jambeck et al., 2015; da Costa, 2017). Approximately 4.8–12.7 million tons of plastic waste have been estimated to enter the ocean in 2010 that exceeds 8 million tons in 2015 and by 2050 this amount is expected to increase to around 32 million tons per year (Jayasiri et al., 2013; Ng and Obbard, 2006). Approximately 33% of all the plastic produced each year is discarded within a 12-month period by considering non-reusable. The discarded plastics as waste enters the aquatic system (including the ocean) by littering or dumping or spillages from land source.

Various legislative and/or regulatory instruments exist to control, manage and reduce the use of plastics. Policy interventions have been implemented in the form of bans, levies, taxes and volunteer efforts through the 3R rule (reduce, reuse and recycle) plastics (UNEP, 2018). But these policy tools have a low impact in the scenario of increased volume production and synthesis of new material with new or enhanced properties. Governments have also implemented various policy instruments to control plastic wastes as an individual effort or collectively.

2.1 LAWS, REGULATIONS OR POLICY TOOLS AT REGIONAL OR NATIONAL LEVELS

2.1.1 LAWS, REGULATIONS OR POLICY TOOLS IN ASIA

Table 2.1 indicates laws, regulations or policy tools in Asia for single-use plastics. In various countries of Asia, the control in the use and manufacture of plastics has been attempted by levies and bans, for example, in Bangladesh. Despite prohibitions and levies, the enforcement of regulations has often been poor, and single-use plastics continue to be mismanaged and widely used. In contrast, Japan is another example where relatively limited leakages of single-use plastics in the environment are attributed to a high degree of social consciousness and effective waste management systems.

TABLE 2.1

Laws, Regulations or Policy Tools about Plastic Material in Asia

Country	Policy/Year/Level	Policy Description and Impact	References
Bangladesh	Ban – in force/2002/ National	Ban on polyethylene plastic bags. Initial positive response from the public. Use of plastic bags increased after some years due to lack of enforcement and absence of cost-effective alternatives	IRIN (2011)
Bhutan	Ban – in force/2009/ National	Ban on plastic bags. Bags are still commonly used. Compliance has been difficult to monitor.	Clean Bhutan (2015)
China	Ban and levy – in force/2008/National	Ban on non-biodegradable plastic bags <25 μm and levy on consumer for thicker ones. In Chinese supermarkets, plastic bag use decreased between 60 and 80%. Ban has been ineffectively enforced in food markets and among small retailers.	Xanthos and Walker (2017)
	Levy – in force /2009 /Local – Hong Kong	Levy on consumer. But implementation in different phases. Initially limited Impact due to implementation only in selected chains and outlets. In 2015, the levy was extended to over 100,000 retailers. 25% fewer bags were disposed in landfills within 1 year.	Kao (2013)
	Ban – in force/2015/ Local – Jilin province	Ban on production and sale of non-biodegradable plastic bags and tableware in Jilin province.	Sun (2015)
India	Ban – in force / 2016 /National	Ban on non-compostable plastic bags <50 μm.	India, Ministry of Environment, Forest and Climate Change, (2016)
	Ban – in force/2004/ Local – Himachal Pradesh	Ban on the production, storage, use, sale and distribution of non-biodegradable plastic bags. In 2011 a ban on disposable plastic products, such as plastic cups, drinking glasses and plates was introduced	Duboise (2012)
	Ban – in force/2016/ Local – Karnataka	Ban on manufacturing and sale of plastic bags in the Indian state of Karnataka. Plastic bags continue to be both available and commonly used.	DHNS (2017); Deepika (2017)
	Ban – in force/2016/ Local – Punjab	Ban on the manufacture, stocking, distribution, sale or use of single-use plastic carry bags and containers in the state of Punjab.	PTI (2016)
	Ban – in force/2010/ Local – Haryana	Ban on manufacture, stocking, distribution, sale or use of plastic carry bags in the state of Haryana	NDTV India (2010)

(Continued)

TABLE 2.1 (Continued)

Laws, Regulations or Policy Tools about Plastic Material in Asia

Country	Policy/Year/Level	Policy Description and Impact	References
	Ban – in force/2016/ Local – Kerala	Ban on plastic bags <50 μm in the Indian state of Kerala	Deccan (2016)
	Ban – in force/2001/ Local – West Bengal	Several regulations from 2001 onwards. Ban on plastic bags <40 μm and blanket ban in certain areas in West Bengal. Plastic bags are still commonly used. Implementation is limited.	Mahesh et al. (2015)
	Ban – in force/1998/ Local – Sikkim	Ban on delivery or purchasing of goods and materials in plastic wrappers or plastic bags in the state of Sikkim. Although plastic bags are still common (used by 34% of shops) the majority switched to paper bags or newspaper (66%).	Bari (2018)
	Ban – in force/2016/ Local – Sikkim	Ban on sale and use of disposable Styrofoam in Sikkim.	Styrofoam ban in Sikkim (2016)
	Ban – in force/2017/ Local – New Delhi	Ban on all kinds of disposable plastics in New Delhi	Naik (2017a); Bari (2018)
	Ban – in force/2018/ Local – Maharashtra	Ban on plastic bags <40 μm in the state of Maharashtra	Naik (2017b)
Indonesia	Levy – in force/2017/ Local – 23 cities	Levy on plastic bags imposed on customers (equivalent to $0.015 per bag) at selected retailers in 23 cities. 40% reduction, on average, in the number of plastic bags used in the selected cities, but resistance has been seen from consumers and the plastic industry. The government is considering the imposition of a nationwide tax on plastic bags starting from 2018.	Black (2016)
	Ban – in force/2017/ Local – Banjarmasin	Ban on plastic bags in the city of Banjarmasin. Reduced bag consumption by 80%.	Jong (2017)
	Ban – in force/2016/ Local – Bandung	Ban on the use of Styrofoam in the city of Bandung.	Hong (2016)

(Continued)

TABLE 2.1 (Continued)

Laws, Regulations or Policy Tools about Plastic Material in Asia

Country	Policy/Year/Level	Policy Description and Impact	References
Israel	Ban and Levy – in force/2017/ National	Ban on bags <20 µm and levy on thicker ones in supermarkets (around $0.03). A survey revealed that, 4 months after the law came into effect, 42% of shoppers had not bought any plastic bags from supermarkets.	Udasin (2016); Raz-Chaimovich (2017)
Malaysia	Levy – in force/2011/ Local – Penang State	MYR 0.20 charge on plastic bags, in line with the campaign: "No free plastic bags"	Zen et al. (2013); The Straits Times (2017)
	Ban – in force/2012/ Local – Penang State	Ban on polystyrene	
	Ban – in force/2017/ Local – Federal Territories	Ban on non-biodegradable plastic bags and food containers in Malaysia's Federal Territories (Kuala Lumpur, Putrajaya and Labuan)	
Mongolia	Ban – in force/2009/ National	Ban on the importation and use of non-biodegradable plastic bags <25 µm. After a few years the ban was incorporated into a new "Waste Law", negatively affecting the enforcement of the ban and the administrative supervision	Zoljargal (2013)
Myanmar	Ban – in force/2009/ Local – Mandalay and Nay Pyi Taw	Ban on the use of small and thin plastic bags in Mandalay and Nay Pyi Taw	People's Daily Online (2009)
	Ban – in force/2011/ Local – Yangon	Ban on the production, storage and sale of polyethylene bags in Yangon	The Independent (2011)
Pakistan	Ban – in force/2013/ Local – Islamabad Capital Territory	Ban on the sale, purchase and use of polyethylene bags in the Islamabad Capital Territory, and introduction of oxo-biodegradable plastic bags	Naeem (2013)
	Ban – in force/2013/ Local – Punjab	Ban on the manufacturing, sale and usage of non-degradable plastic products in Punjab.	Masud (2017)
	Ban – in force /2018/ Local – Sindh	Ban on certain non-degradable plastic products, including carrier bags, in Sindh. (The Sindh Prohibition of Non-degradable Plastic Products.	The Express Tribune (2018)

(Continued)

TABLE 2.1 (*Continued*)

Laws, Regulations or Policy Tools about Plastic Material in Asia

Country	Policy/Year/Level	Policy Description and Impact	References
	Ban – in force/2017/ Local – Khyber Pakhtunkhwa	Ban on the manufacture, importation, sale and use of non-biodegradable plastic bags and regulation of oxo-biodegradable plastic products in Khyber Pakhtunkhwa	Khattak (2017)
Philippines	Ban – in force/2011/ Local – Muntinlupa	Ban on the production, importation and distribution of single-use plastic bags in major supermarkets and levy on consumers on thicker ones (>50 µm)	Quillen (2017); Martina (2017)
Sri Lanka	Ban – in force/2017/ National	Ban on the import, sale and use of polyethylene bags	Agence France-Presse (2017); Jayasekara (2017)
Viet Nam	levy – in force/2012/ National	Non-biodegradable plastic bags are taxed by weight at VND 40,000 ($1.76) per kilogram (levy on retailer). Plastic bags are still widely used across Viet Nam. The government is considering an amendment to increase the tax fivefold.	Tuoi Tre News (2017)
South Korea	Ban – in force/2019	South Korean Ministry of Environment announces that 2,000 discount chain outlets and 11,000 supermarkets could be fined up to three million won if they ignore the ban.	https://resource.co/article/south-korea-latest-country-ban-single-use-plastic-bags-13028

2.1.2   Laws, Regulations or Policy Tools in Africa

As indicated in Table 2.2, several countries of Africa instituted a total ban on the production and use of plastics particularly plastic bags that was introduced from 2006 to 2018 and shifted into implementation (approximately 58%) from 2014 to 2017 (UNEP, 2018).

2.1.3   Laws, Regulations or Policy Tools in Europe and Oceania

In Europe, countries choose measures ranging from bans, levies and laws to agreements with the private sector, in response to EU Directive 2015/720, to achieve a sustained reduction in the number of lightweight plastic bags used per person by 2025 (Tables 2.3 and 2.4). Currently, the European Commission is finalizing a "European Strategy for Plastics in a Circular Economy" (2018–2030) to eliminate over-packaging and reduce unnecessary generation of single-use plastic wastes (UNEP, 2018; EC, 2013).

As mentioned in Table 2.4, most of the Australian states have banned non-biodegradable and lightweight plastic bags (UNEP, 2018) as a measure to control or limit plastic wastes.

2.1.4   Laws, Regulations or Policy Tools in Central, South and North America

Mostly in Central and South America, the policy instruments and regulations are in place to control the consumption of plastic bags. Countries such as Haiti and Costa Rica have effectively regulated the use of foamed plastics (UNDP, 2017). Some laws, regulations and policy tools, as national or subnational efforts, are listed in Table 2.5.

As represented in Table 2.6, in North America, rules and regulations on lightweight plastic bags have been introduced in the form of bans to achieve sustained reduction, for example, in Montreal, California and Hawaii. Action has also been taken against single-use styrofoam products in New York City, which reinstated in 2017 its ban after a first attempt in 2015 (Babin, 2017).

2.2   LAWS, REGULATIONS OR POLICY TOOLS AS COLLECTIVE EFFORTS

In prospects of plastic pollution control, the solutions are complex, multilateral and transboundary. So the global response must be holistic and dynamic, requiring coordinated action by diverse stakeholders at the regional, national, and international levels. Despite the growing concern for long-term, multilateral and comprehensive action, no firm international agreements exist those focus primarily on combating all plastic pollution. Some of the attempts have been made to address all types of the plastic crisis on a global scale, from wellhead extraction to ocean pollution.

TABLE 2.2

Laws, Regulations or Policy Tools about Material in Africa

Country	Policy/Year/ Level	Policy Description and Impact	References
Benin	Ban – in force/2018/ National	Ban on import, production, sale and use of non-biodegradable plastic bags	LégiBénin (2018)
Botswana	Levy – in force/2007/ National	Levy on retailer. No enforcement upon retailers to charge for plastic bags. Retailers decide if and how much to charge. As an impact the tax resulted in voluntary levy on plastic bags. Decline in the consumption of plastic bags: 50% drop within 18 months, partial success probably due to consistently high prices of bags. However control over pollution resulting from plastic carrier bags failed, leading to discussions about banning them.	Dikgang and Martine (2010)
Burkina Faso	Ban – in force/2015/ National	Ban on production, import, marketing and distribution of non-biodegradable plastic bags.	UNDP-UNEP (2015)
Cameroon	Ban – in force/2014/ National	Ban on non-biodegradable plastic bags. Due to a lack of inexpensive alternatives, plastic bags appear to be smuggled from neighboring countries. The government has tried to encourage the cleanup of plastic litter by paying citizens for each kilogram of plastic waste collected. Through the program, an estimated 100,000 kg of plastic waste was collected in 2015 alone	Nformgwa (2014); Colbert (2016)
Cape Verde	Ban – in force/2017/ National	Ban on the sale and use of plastic bags.	EnviroNews Nigeria (2017)
Chad	Ban – in force/2010/ Local – N'Djamena	Ban on the importation, sale and use of plastic bags in the capital city, N'Djamena. Less observable plastic pollution in the city.	IRIN (2010)

(Continued)

TABLE 2.2 (*Continued*)
Laws, Regulations or Policy Tools about Material in Africa

Country	Policy/Year/Level	Policy Description and Impact	References
Côte d'Ivoire	Ban – in force/2014/ National	Ban on the importation, production, use and sale of non-biodegradable plastic bags <50 μm.	Boisvert (2014)
East Africa	Ban – in force/2017/ Regional	The East African Legislative Assembly (EALA) introduced a ban on the manufacturing, sale, importation and use of polythene bags under the East African Community Polythene Materials Control Bill 2017.	Karuhanga (2017)
Egypt	Ban – in force/2009/ Local-Hurghada	Ban on the use of plastic bags in Hurghada. Distribution of 50,000 cloth bags for free by the Hurghada Environmental Protection and Conservation Association, together with letters explaining the health and environmental reasons behind the campaign.	Zohny (2009)
Eritrea	Ban – in force/2005/ National	Ban on the importation, production, sale and distribution of plastic bags. Problems associated with plastic bags, such as the blockage of drains and water pipes, dramatically decreased	Fikrejesus (2017)
Ethiopia	Ban – in force/2007/ National	Ban on production and importation of non-biodegradable plastic bags <30 μm	Ethiopian News Agency (2016); Sisay (2016); Alicia (2011)
Gambia	Ban – in force/2015/ National	Ban on the sale, importation and use of plastic bags. Success was seen in the first phase after implementation, but there has been a reappearance after a political impasse	Coker (2017)
Guinea-Bissau	Ban – in force/2016/ National	Ban on the use of plastic bags. But Law was not strictly enforced. Strong resistance from both consumers and retailers, claiming a lack of consultation	Guinea-Bissau (2017)
Kenya	Ban – in force/2017/ National	Ban on the importation, production, sale and use of plastic bags	The Guardian (2017)

(Continued)

TABLE 2.2 (*Continued*)
Laws, Regulations or Policy Tools about Material in Africa

Country	Policy/Year/ Level	Policy Description and Impact	References
Malawi	Ban – in force/2015/ National	Ban on the use, sale, production, exportation and importation of plastic bags <60 μm.	UNDP (2015)
Mali	Ban – approved/2012/ National	Ban on the production, importation, possession, sale and use of non-biodegradable plastic bags.	Braun and Traore (2015)
Mauritania	Ban – in force/2013/ National	Ban on the manufacture, use and importation of plastic bags.	BBC News (2013)
Morocco	Ban – in force/2009/ National	Ban on the production, importation, sale and distribution of black plastic bags. Although only considered partially successful, the law is considered an important step forward.	Ellis (2016); UNEP (2018)
	Ban – in force/2016/ National	Ban on the production, importation, sale and distribution of plastic bags. 421 tons of plastic bags were seized in 1 year. Citizens have switched to fabric bags. The Moroccan government declared that plastic bags are virtually no longer used in the country	
Mozambique	Ban – in force/2016/ National	Ban on the production, importation, possession and use of plastic bags <30 μm. People were advised to use baskets made from either grass or coconut trees	Mozambique News Agency (2015); Zitamar News (2016)
Niger	Enacted legislation, 2014, Ban – in force/2015/ National	Ban on production, importation, usage and stocking of plastic bags. But the impact is limited due to pure enforcement.	Panapress (2014); UNEP (2018)

(Continued)

TABLE 2.2 (Continued)

Laws, Regulations or Policy Tools about Material in Africa

Country	Policy/Year/Level	Policy Description and Impact	References
Nigeria	Ban – in force/2013/National	Ban on production, importation, usage and stocking of low-density smooth plastic and packaging bags.	Obateru (2016)
Rwanda	Ban – in force/2008/National	Ban on the production, use, importation and sale of all polyethylene bags. In the first phase the ban resulted in a black market for plastic bags. Over time, plastic bags were replaced by paper bags.	Clavel (2014); Pilgrim (2015)
Republic of the Congo	Ban – in force/2011/National	The government announced a ban on the production, importation, sale and use of plastic bags, but did not announce when it would take effect.	Reuters (2011)
Senegal	Ban – in force/2016/National	Ban on the production, importation, possession and use of plastic bags <30 μm	Associated Press (2015); UNEP (2018)
Somalia	Ban – in force/2015/National	Ban on disposable plastic bags in Somaliland. Despite the law, plastic bags are still widely used	Masai (2015); Hasan (2017)
South Africa	Ban – in force/2003/National	Ban on plastic bags <30 μm and levy on retailer for thicker ones. In the first phase the consumption of plastic bags fell, but then increased again due to lack of enforcement.	Dikgang et al. (2012)
Tanzania	Ban – approved/2006/National	Ban on plastic bags and bottles. But ban has not been implemented. Since then, the government made constant efforts to phase out plastic bags. The latest ban was issued in 2016, but implementation has been pushed back	Kiprop (2017)
	Ban – in force/2006/National	Ban on the importation, distribution and sale of plastic bags <30 μm.	

(*Continued*)

TABLE 2.2 (*Continued*)

Laws, Regulations or Policy Tools about Material in Africa

Country	Policy/Year/ Level	Policy Description and Impact	References
Tunisia	Ban – in force/2017/ National	Ban on the production, importation and distribution of single-use plastic bags in major supermarkets and levy on consumers on thicker ones (>50 μm)	Quillen (2017)
Uganda	Ban – in force/2009/ National	Ban on lightweight plastic bags <30 μm. Enforcement was weakened by manufacturers' lobby. Implementation attempts by the National Environmental Management Agency (NEMA) in April 2015 had no sustained impacts. Plastic bags can still be found in some parts of the country, although local entrepreneurs started to produce woven, reusable bags	Wakabi (2015); Namara (2016)
Zimbabwe	Ban and levy – in force/2010/ National	Ban on plastic bags <30 μm and levy on consumer for thicker ones. Implementation has been difficult because of resistance from the informal sector	Chitotombe and Gukurume (2014)
	Ban – in force/2017/ National	Ban on Styrofoam products. Ban temporary lifted shortly after its introduction to allow businesses more time to replace Styrofoam containers with recyclable or biodegradable ones	Mhofu (2017)
Mauritius	Ban – in force/2016/ National	Ban on the importation, manufacture, sale or supply of plastic bags, with 11 Types of plastic bags for essential uses and hygienic and sanitary purposes exempt (e.g., roll-on bag for meat products, waste disposal bags, bags as integral part of packaging, bags manufactured for export)	http://www.govmu. org/English/News/Pages/-Mauritius-bans-the-use-of-plastic-bags.aspx.

TABLE 2.3

Laws, Regulations or Policy Tools about Plastic material in Europe

Country	Policy/Year/Level	Policy Description and Impact	References
European Union		EU directive: Member states must ensure that by the end of 2019 no more than 90 lightweight (<50 μm) bags are consumed per person per year. By the end of 2025 that number should be down to no more than 40 bags per person. Member states can choose whether to introduce bans, taxes or other policy tools (European Union, 2015)	
Netherlands	Levy – in force/ 2016/National	Levy on consumer. Very lightweight bags for primary packaging are exempt. While businesses have the freedom to decide how much they will charge, the official guideline is €0.25 per bag (around $0.30)	Pieters (2017)
Ireland	Levy – in force/ 2002 (revised in 2007)	Levy on consumer for plastic bags (initially, €0.15, later, €0.22); Legislation allows the levy to be amended, not exceeding €0.70 per bag. one year later, the consumption of plastic bags decreased by more than 90%	O'Neill (2016)
Belgium	Levy – in force/ 2007/National	Levy on consumer. A bill on carrier bags has been drafted but is yet to be adopted. Consumption of carrier bags decreased by 80% over the following decade.	Surfrider Foundation Europe (2017); Alpagro (2016)
	Ban – in force/ 2016/Local-Wallonia	Ban on the use of single-use plastic bags in Wallonia. Exception of thin compostable bags for foods that can be moist, until the end of 2018	Surfrider Foundation Europe (2017)
	Ban – in force/ 2017/Local –Brussels Capital Region	Ban on non-compostable plastic bags <50 μm in the Brussels Capital Region	Alpagro (2016)
Bulgaria	Levy – in force/2011/National	Levy on supplier of PE bags (<15 μm). Drastic reduction in the use of plastic bags according to the Bulgarian Ministry of Environment and Water.	Surfrider Foundation Europe (2017); Bulgaria's Environment Ministry Reports (2015)
Croatia	Levy – in force/2014/National	Levy on supplier, with levies to go to the Environmental Protection and Energy Efficiency Fund	Surfrider Foundation Europe (2017)
Cyprus	Levy – approved/2018/ National	Levy on consumer (€ 0.05, around $0.06) for plastic bags in supermarkets.	CNA News Service (2018)

(Continued)

TABLE 2.3 (Continued)

Laws, Regulations or Policy Tools about Plastic material in Europe

Country	Policy/Year/Level	Policy Description and Impact	References
Czech Republic	Levy – approved/2018/ National	Levy on consumer for plastic bags >15 μm. Retailers determine the price, but charge must at a minimum cover the production cost of the plastic bag.	Plastic Portal (2018); News Expats (2017)
Denmark	Levy – in force/1994/National	Levy on supplier for plastic bags. Fee passed on to retailers, who in turn pass it on to consumers (currently a bag costs around $0.56 per bag). A decrease from around 800 million bags to half of that.	Larsen (2014)
Latvia	Levy – in force/2009/National	Levy on retailer for plastic carrier bags. Most supermarkets charge for plastic carrier bags. Plastic bag consumption dropped. The use of reusable bags increased, but stabilized after the first year	Brizga (2016)
Lithuania	Levy – approved/2009/ National	Levy on consumer. Prohibition of free lightweight plastic bags with a thickness between 15 and 50 μm. Supposed to enter into effect by 31 December 2018	Lithuania, Ministry of the Environment (2016); Surfrider Foundation Europe (2017)
Malta	Levy – in force/2009/ National	Levy on consumer for all plastic bags (€0.15).	Xuereb (2009)
Romania	Levy – in force/2009, law – approved/2018 / National	Ban on plastic bags <50 μm in supermarkets and 15 μm on national markets.	Marica (2018)
Italy	Ban – in force/2011/National	Ban on non-biodegradable plastic bags. Reduction of plastic bag consumption by approximately 55%.	Surfrider Foundation Europe (2017); UNEP (2018)
	Levy – in force/2018	Levy on consumer for plastic bags in supermarkets and grocery stores. Only biodegradable and compostable lightweight bags are allowed to be provided or sold.	
Hungary	Levy – in force/2012/National	Levy on supplier: The introduction of the fee obliged producers and distributors to pay the fee, which was incorporated into the products' price. Retailers voluntarily added a fee on plastic bags.	EC (2013); Kis (2015)

(Continued)

TABLE 2.3 (*Continued*)
Laws, Regulations or Policy Tools about Plastic material in Europe

Country	Policy/Year/Level	Policy Description and Impact	References
United Kingdom	Levy – in force/2013/ Local – Northern Ireland	Levy on consumer for plastic bags (£0.05, around €0.06). The first year saw a drop of 71% in the consumption of plastic bags. The second year showed an added 42.6% decrease.	BBC News (2015)
	Levy – in force/2014/ Local – Scotland	Levy on consumer (£0.05, around €0.06). Plastic bag usage declined by 80% in 1 year.	BBC News (2015)
	Levy – in force/2015/ Local – England	Levy on consumer (£0.05, around €0.06) for plastic bags. Charged by companies with >250 employees. Voluntary basis for smaller retailers. Single-use plastic bags used dropped by more than 85% in the 6 months.	Smithers (2016)
Portugal	Levy – in force / 2015 / National	Levy on supplier, but the fee (€0.10) was largely passed onto the consumer. Following implementation, consumption of lightweight plastic bags decreased by 74%. Consumption of reusable plastic bags, exempted from the levy, increased by 61%.	Martinho et al. (2017)
France	Ban – approved/2015/ National	Ban on all disposable tableware not made from at least 50% biologically derived by 2020.	Eastaugh (2016); CNN (2016)
	Ban – in force/2016/ National	Ban on lightweight single-use carrier bags (<50 μm and <10L). Expanded in 2017 to all other plastic bags except compostable bags.	EuroNews (2016)
Sweden	Law – in force / 2017 / National	Requires supermarkets to educate customers on the environmental effects of plastic bags.	Harford (2017)
	Levy – in force / 2011 / Local – Wales	Levy on consumer for plastic bags (£0.05) in Wales. The consumption of single-use plastic bags has declined by more than 70% since the tax was introduced.	Morris (2015)
Greece	Levy – in force / 2018 / National	Levy on consumer (€0.04) for non-biodegradable plastic bags (<50 μm). Businesses allowed to charge customers for thicker bags (up to 70 μm). After the first month of implementation lightweight plastic bag consumption decreased by 75%–80% and sales of reusable shopping bags increased sharply	Manifava (2017); Smith (2018)

(*Continued*)

TABLE 2.3 (*Continued*)

Laws, Regulations or Policy Tools about Plastic material in Europe

Country	Policy/Year/Level	Policy Description and Impact	References
Slovakia	Levy – in force / 2018 / National	Levy on consumer for plastic bags between 15 and 50 μm.	Plastic Portal (2018)
Poland	Levy – in force / 2018 / National	The government is planning the introduction of a PLN 1 (around $0.28) levy on plastic bags Implementing measures were notified to the European Commission in November 2017.	Surfrider Foundation Europe (2017)
Spain	Levy in 2017 and Draft law approved/2020	A tax of €0.45 per kilogram of plastic waste. Also includes a ban on straws. To be levied on the manufacturing, import or intra-EU acquisition of the non-reusable plastic packaging	Gerrard (2018)
Estonia	Levy – in force / 2017 / National	Levy on consumer on plastic bags (<50 μm (exemption of very lightweight bags used for hygiene and prevent food waste). Avoidance of sale or free of charge oxo-degradable plastic carrier bags.	Larsen & Venkova (2014); https://www.riigiteataja.ee/en/eli/ee/512012016003/consolide/current

TABLE 2.4

Laws, Regulations or Policy Tools about Plastic Material in Oceania

Country	Policy/Year/Level	Policy Description and Impact	References
Fiji	Levy – in force/2017/ National	Levy on consumer, FJD 0.10 ($0.05) per plastic bags.	https://www.loopvanuatu.com/vanuatu-news/vanuatu-joins-other-pics-tackle-plastic-problems-64061
Papua New Guinea	Ban – in force/2016/ National	Ban on non-biodegradable plastic shopping bags.	Wayang (2017)
Vanuatu	Ban – in force/2018/ National	Ban on manufacture, use and import of single-use plastic bags, straws and polystyrene takeaway food containers. Bags to wrap and carry fish or meat are exempt.	SPREP (2018)
Marshall Islands	Ban – in force/2017/ National	Ban on importation, manufacture and use of single-use plastic carrier bags. Ban on Styrofoam and plastic cups, plates and packages	Styrofoam and Plastic Products Prohibition Act (2016)
Palau	Ban – in force/2017/ National Islands	Ban on the importation and distribution of plastic shopping bags	Carreon (2017); SPREP (2018)
Australia	Ban – in force /2003/ Local – Coles Bay	Ban on non-biodegradable plastic checkout bags. It has been estimated that in 10 years, the ban has avoided the use of two million plastic bags.	Fickling (2003); Twomey (2013)
	Ban – in force/2009/ Local – South Australia	Ban on lightweight plastic bags in South Australia. Consumption of reusable, thicker plastic bags increased.	Watson (2013)
	Ban – in force/2011/ Local – Australian Capital Territory	Ban on lightweight plastic bags. Two years after the implementation of the ban, 36% reduction in the amount of plastic bag waste in landfills.	Hayne (2017)
	Ban – in force/2011/ Local – Northern Territory	Ban on plastic bags <35 μm. A survey revealed that, 5 years after the ban was introduced, plastic bag litter increased .	Rigby et al. (2017)
	Ban – in force/2013/ Local – Tasmania	Ban on plastic bags <35 μm. As an impact an increased consumption of thicker bags was observed.	Richards (2017)
	Ban approved/2018/ Local – Queensland	Ban on plastic bags <35 μm in Queensland.	Cooper (2017a, b)
New Zealand	Levy – in force/2017/ Local	Levy on plastic bags. Almost half of the nation's mayors have signed an open letter to the Ministry of the Environment to impose a mandatory charge on plastic bags. A supermarket chain launched a campaign, letting shoppers decide how much to pay (or not) for plastic bags. Another supermarket chain announced that it will phase out all plastic bags by 2018.	Cann (2017); Huffadine (2017); Clayton (2017)

TABLE 2.5

Laws, Regulations or Policy Tools about Plastic Material in Central and South America

Country	Policy/Year/Level	Policy Description and Impact	References
Antigua and Barbuda	Ban – in force/2016/ National	Ban on the use and importation of plastic bags	https://elaw. org/plastic/AG_AntiguaBarbuda_PlasticLaws
	Ban – in force/2017/ National	Ban on Styrofoam with an implementation plan of three stages. Ban on food service containers since 2017, from 2018 onwards ban on plastic utensils (e.g., spoons, straws, food trays, etc.) and ban on importation and use of Styrofoam coolers	Antigua Nice Ltd (2017)
Argentina	Ban – in force/2017/ Local – Buenos Aires	Ban on non-biodegradable plastic shopping bags <50 μm in Buenos Aires. Shortly after the ban was introduced, sales of changuitos (individual shopping carts) rose sharply.	Martina (2017); Tavella (2017)
	Ban – in force/2009/ Local – Córdoba	Ban on the use of polyethylene bags in Córdoba	
Belize	Ban – in force/2008/ National	Ban on single-use plastic shopping bags, Styrofoam and plastic food utensils.	Belize Press Office (2018)
Brazil	Levy–in force /2009/ Local – Rio de Janeiro	Requirement to substitute polyethylene and polypropylene bags with alternatives, or, if not done, to take back any quantity of plastic bags from any source and dispose of them properly and compensate the public by giving them a discount if they bring their own bag, or to pay them with food products for every 50 plastic bags they bring. Reduction of 24% of plastic bags used each year.	Siqueira (2011)
	Ban – in force/2015/ Local – Sao Paulo	Ban on non-biodegradable plastic bags in Sao Paulo	Petrone (2015)
Ecuador	Ban – in force/2015/ Local – Galápagos Islands	Ban on plastic bags in the Galápagos Islands	Haskell (2014)

(Continued)

(Continued)

TABLE 2.5 (Continued)

Laws, Regulations or Policy Tools about Plastic Material in Central and South America

Country	Policy/Year/Level	Policy Description and Impact	References
Colombia	Ban and levy – in force /2017/ National	Ban on disposable plastic bags smaller than 30×30cm and levy on consumer on single-use plastic bags (20 Colombian pesos, around $1). 27% reduction in the use of plastic bags.	UNEP (2017)
Guatemala	Ban – in force/2017/ Local – San Pedro La Laguna and other cities	Ban on plastic bags and Styrofoam containers in San Pedro La Laguna. Cantel, Quetzaltenango and San Juan Sacatepéquez have introduced similar laws.	Chiyal (2017)
Haiti	Ban – in force/2013/ National	Ban on the importation and production of plastic bags and Styrofoam containers.	Lall (2013)
Honduras	Ban – in force/2016/ Local – Roatán, Utila, Guanaja	Ban on plastic bags instituted at the municipal level in Roatán, Utila and Guanaja. Accompanied by an awareness-raising campaign. 100% elimination in Guanaja, 80% decline on Utila and 50% decline in Roatán.	The Summit Foundation (2017)
Mexico	Ban and levy – in force/2010/ National	Retailers in Mexico City must charge for plastic bags, which, according to the law, must also be biodegradable	Malkin (2009)
	Ban – in force/2018/ National	Ban on disposable plastic bags in Queretaro City	Reyes (2017)
Panama	Ban – in force/2018/ National	Ban on the sale and use of non-biodegradable plastic bags	https://www.livekindly. co/panama-first-central-american-country-ban-plastic-bags
St. Vincent and the Grenadines	Ban – in force/2017/ National	Ban on the importation of Styrofoam products used for sale or storage of food; value-added tax (VAT) removed from biodegradable alternatives to lower their cost.	https://oceanconference. un.org/commitments/?id=18100
Costa Rica	Ban – in force/2021/ National	The government announced the phasing out of all kinds of disposable plastics by 2021.	UNDP (2017)

TABLE 2.5 (Continued)

Laws, Regulations or Policy Tools about Plastic Material in Central and South America

Country	Policy/Year/Level	Policy Description and Impact	References
Jamaica	Ban – in force/2018/ National	The government is considering the introduction of a ban on non-biodegradable plastic bags below 50-gallon capacity and on Styrofoam containers.	Serju (2017)
Uruguay	Ban – in force/2017/ National	Levy on consumer on single-use plastic bags	Lu (2016)
Chile	Ban – in force/2014/ Local – Punta Arenas	Ban on polyethylene bags except for perishable food products.	http://www.patagonjournal.com/index.php?option=com_content&view=article&id=4152%3Achiles-plastic-bag-ban-opens-possibilities-for-the-uture&catid=78%3Amedioambiente&Itemid=268&lang=en; UNEP (2018)
	Bill approved/2017/ National	Ban on the sale of plastic bags in 102 coastal villages and towns	Chow (2017)
Guyana	Ban – in force/2016/ National	Ban on the importation and use of Styrofoam items.	https://parliament.gov.gy/documents/-documents-laid/5543-enivornmental_protection_(expanded_polystyrene_ban)_regulations_no._8_of_2015.pdf

TABLE 2.6

Laws, Regulations or Policy Tools about Plastic Material in North America

Country	Policy/Year/Level	Policy Description and Impact	References
United States of America	Levy – in force/2000/ Local – Washington, DC	Levy on consumer for plastic bags ($0.05) in Washington, DC. A survey in 2014 revealed that the consumption of plastic bags decreased on average from 10 to 4 plastic bags a week.	UNEP (2018); https://doee.dc.gov/bags
	Ban – in force/2011/ Local – American Samoa	Ban on the sale and use of petroleum-based plastic bags (some exceptions possible for fresh and frozen products and others)	American Samoa Environmental Protection Agency (2011)
	Ban – in force/2011/ Local – Hawaii	Ban on single-use plastic bags in Hawaii. 2013: Big Island Hawaii, 2018: Honolulu (ban and fee), 2011: Kauai, 2008: Maui and Pala	UNEP (2018); https://www.hawaiizerowaste.org/reuse-2/county-of-hawaii-plastic-bag-reduction-ordinance-2/
	Ban and levy – in force/2012/ Local – San Francisco, California	Ban on single-use checkout plastic bags and levy on consumer on compostable bags, recycled paper bags or reusable (>125 uses) bag of $0.10 in the county and city of San Francisco	UNEP (2018)
	Ban – in force/2013/ Local – Austin, Texas	Ban on single-use plastic bags (<101 μm). While the consumption of single-use plastic bags decreased, that of reusable, thicker plastic bags increased	Richards (2017)
	Ban – in force/2015/ Local – New York City, New York	Ban on single-use Styrofoam containers instituted in New York City. The ban was challenged by a coalition of recycling firms and plastics manufacturers who claimed the material is recyclable. The ban was lifted in 2015 and reintroduced in 2017.	Alexander (2017)
	Ban – in force/2016/ Local – California	Ban on single-use plastic bags and levy on thicker reusable ones (US$ 0.10) in California. Plastic bags accounted for about 3% of the litter collected during the 2017 Coastal Cleanup Day, compared to 7.4% in 2010	Los Angeles Times Editorial Board (2017)

(Continued)

TABLE 2.6 (*Continued*)

Laws, Regulations or Policy Tools about Plastic Material in North America

Country	Policy/Year/Level	Policy Description and Impact	References
	Levy – in force/2017/ Local – Chicago, Illinois	Levy on consumer plastic bags in Chicago ($0.07). The number of plastic bags (and paper bags, as these are also taxed) declined by 42% one month after the introduction of the tax.	Cherone and Wetli (2017)
	Ban – in force/2017/ Local – Seattle	Ban on single-use plastic bags, including bags labeled with biodegradable, degradable, decomposable or similar, and voluntary levy on thicker (>57 μm) plastic. bags	Seattle, Washington, USA (2017)
Canada	Ban – in force/2007/ Local – Leaf Rapids	Ban on plastic bags	Duboise (2012)
	Ban – in force/2010/ Local – Wood Buffalo	Ban on single-use plastic bags (>57 μm)	Wood Buffalo (2010)
	Levy–in force /2010/ Local – Thompson	Ban on the sale or giveaway for free of plastic shopping bags in Thompson	Duboise (2010)
	Ban – in force/2018/ Local – Montreal	Ban on plastic bags <50 μm.	Fundira (2016)

2.2.1   UNITED NATIONS CONVENTION ON THE LAW OF THE SEA

United Nations Convention on the Law of the Sea (UNCLOS) is perhaps the most widely known, colloquially described, regulatory tool available worldwide (Naik, 2017b). UNCLOS came into force on 16 November 1994 after being opened for signature in December 1982. UNCLOS constituted an unparalleled attempt to regulate all aspects of the uses of the ocean and resources of the sea, and thus bring a stable order to very source of life, as mentioned in the signed document. UNCLOS that is composed of 320 articles focuses on an extensive array of subjects, ranging from economic and territorial jurisdiction, legal status of resources on the seabed beyond national jurisdiction limits, to navigational rights and binding procedures for settlement of disputes among states. It also applies to marine resources conservation and management as well as preservation and protection of the marine environment, to which 46 articles are devoted (Articles 192–237, Part XII). Article 210, for example, mandates that all signatory states must develop frameworks to prevent, reduce and control pollution of the marine environment by dumping. Concurrently, any state can have the right to regulate, permit and control such dumping after due consideration of the matter with other states because their geographical situation may be adversely affected (Naik, 2017b). However, the detailed principles and measures foreseen in UNCLOS are of limited efficacy due to the matter of fact that plastic litter is not circumscribed to national jurisdiction and the sources of marine debris are difficult to identify. This is further complicated by inherent limitations that derive from historic regional and economic conflicts, such as the Aegean dispute in which Turkey challenges the extension of the Greek territorial waters foreseen in UNCLOS. Additionally, the United States of America, a pivotal regional player in environmental protection and maritime security as well as a major producer of this type of wastes, is not a signatory state (Zen et al., 2013). Perhaps more glaringly, non-compliance not fulfill their responsibilities. Frequently owing to grievances stemming from the added duties that coastal states incur in, among others, search and rescue operations, pollution prevention and remediation and the need for international navigation information systems and infrastructures, whose compensation is not envisioned in UNCLOS. Nevertheless, UNCLOS established a source of dialogue and communication between signatory states and served to initiate a process that in time may actively contribute to cooperative efforts between states aiming for the reduction of plastic litter in the environment (UNEP, 2018).

2.2.2   NATIONAL OCEANIC AND ATMOSPHERIC ADMINISTRATION
AND THE UNITED NATIONS ENVIRONMENT PROGRAMME

The Marine Debris Program of the U.S. National Oceanic and Atmospheric Administration and the United Nations Environment Programme (UNEP) jointly developed a global agenda specifically aimed at prevention, reduction and management of marine debris. Known as the Honolulu Strategy, it is a comprehensive and global collaborative effort to reduce the ecological, and economic impacts of marine debris worldwide and limit plastics pollution effect on human health. This collaborative framework is organized by a set of goals and strategies applicable all over the world, regardless of specific conditions or challenges (Masud, 2017). Nevertheless, due to its non-binding nature, the Honolulu Strategy does not supersede or supplant national, industrial, municipal or international organizational activities and is restricted to participating states and stakeholders. Rather, it provides a central point for efficient coordination and higher degrees of collaboration between all interested parties concerned with marine debris. Active and voluntary participation at multiple levels – international, regional, national and local – from stakeholders within the government, intergovernmental organizations, the private sector and the entire spectrum of civil society is a prerequisite for the successful implementation of the goals described in Honolulu Strategy. This result-oriented framework comprises the following three distinct goals:

1. Reduce the volume production and the impacts of land-based sources of marine debris introduced into the sea.
2. Reduce the amounts and the impacts of sea-based sources of marine debris, including solid waste, abandoned, lost, lost cargo, or discarded fishing gear, and abandoned vessels.
3. Reduce the amounts and the impacts of accumulated marine debris on shorelines, in benthic habitats and in pelagic waters.

2.2.3 International Convention for the Prevention of Pollution from Ships

The International Convention for the Prevention of Pollution from Ships (MARPOL (73/78)) (revised) was developed by the International Maritime Organization. MARPOL has been updated by amendments through the years. The Convention includes regulations aimed at preventing and minimizing pollution from ships – both accidental pollution and that from routine operations – and currently includes six technical Annexes with the objective of reducing pollution of the seas and oceans, including dumping, oil and air pollution, etc. It deals with different types of garbage and specifies the distances from land and the manner in which they may be disposed of and the most important feature of the Annex is the complete ban imposed on the disposal of all forms of plastics into the sea (UNEP, 2018).

Complementary to MARPOL 73/78, guidelines have been formulated by the International Oceanographic Commission and the Food and Agriculture Organization (Martin, 2009) for monitoring of marine litter, as well as on lost, abandoned or discarded fishing gear. Although the flag states have the authority to enforce restrictions on marine pollution in international waters they either lack the resources, the will to fulfill their duty or both (EC, 2013). Additional efforts are required to help close such enforcement gaps and extend the ability of the Convention to achieve the vital goal of protecting the marine environment, which also includes expanding coastal and port state authority and extending the regulatory requirements to track cargo, including oil, from "cradle to grave" in smaller vessels. It is possible, however, that other multi or transnational agreements, such as Free Trade Agreements, could heighten MARPOL 73/78 compliance, through active public participation in trade and dispute resolutions (Larsen and Venkova, 2014).

2.2.4 UNEP's DRAFT RESOLUTION ON MARINE LITTER AND MICROPLASTICS

In 2017, UNEP's Environment Assembly passed a draft resolution, particularly dealing with marine litter and microplastics after being gathered in Nairobi (Kenya) (EC, 2013). In general, this document recognizes the existence of multiple challenges related to the increased production and consumption of plastics. Furthermore, it acknowledged the participation and initiatives of both public and private entities and urges all stakeholders and countries to reduce plastic use and promote environmentally sound alternatives (EC, 2013). In light of this call to action, some cross-industry agreements have been reached and some enterprises have also independently developed efforts in this direction (UNEP, 2018).

2.2.5 EURATEX

The EURATEX, a textile confederation representing around 160,000 companies, exemplified, in Europe, by agreement to prevent the release of microplastics via washing of synthetic textiles into the aquatic environment (EC, 2013). Also, some internationally recognized companies are developing efforts toward reducing their emissions by reducing the use of plastic in their products (planned by Unilever), phasing out single-use plastics (IKEA) (Xuereb, 2009; EC, 2013) and actively replacing plastic products with, e.g., refillable recipients, as is the case in some McDonald's restaurants (Larsen, 2014).

2.2.6 UNITED NATIONS DECADE OF OCEAN SCIENCE FOR SUSTAINABLE DEVELOPMENT

Also in 2017, the United Nations proclaimed the Decade of Ocean Science for Sustainable Development (2021–2030) (Harvey, 2017). This initiative covers broader goals for fighting plastic pollution, in particular, plastic litter. It focuses on the creation and fostering of active interfaces of science and policy aiming at enabling and boosting sustainable management of coastal areas and oceans. Although it is still in its preparatory phase (2018–2020) but under the agenda "the science we need for the ocean we want", this process is motivated by the will to reverse the cycle that declines the health of the oceans. The current increasing awareness and goodwill toward the protection of the oceans and the development of adequate science-based policies constitute a unique opportunity that may culminate in the creation of measures efficiently directed to the preservation of the marine environment. The issue of plastic pollution has also been addressed from a more economically intergovernmental perspective like G7 (Group of Seven) and G20 (Group of Twenty) have devised specific action plans (Wayang, 2017; UNEP, 2018). These emphasize the need to promote resource efficiency, waste reduction and sustainable waste management. However, most of the reported achievements are reduced to workshops, which have, nonetheless, highlighted the need to identify improved solutions for dealing with exact problem and find a sustainable solution.

2.2.7   CONVENTION FOR THE PROTECTION OF THE MARINE ENVIRONMENT OF THE NORTH-EAST ATLANTIC

OSPAR, The Convention for the Protection of the Marine Environment of the North-East Atlantic, is another statutory instrument, signed and ratified by the EU and 15 states. The name of the Convention reflects the merger (and update) of the 1972 Oslo Convention and the 1974 Paris Convention (Oslo and Paris). OSPAR, with an ultimate goal of protecting the environment, promotes and regulates cooperation. From this initiative, specific guidelines for monitoring marine litter on beaches have been developed, which include practical advice, standardized methodologies for accurate quantification and identification of litter and photographic guides. OSPAR contains a series of Annexes, each for a specific area as follows:

1. Prevention and elimination of pollution by dumping or incineration.
2. Prevention and elimination of pollution from offshore sources.
3. Assessment of the quality of the marine environment.
4. Protection and conservation of the ecosystems and biological diversity of the maritime area (UNEP, 2018).

2.2.8   UNEP's REGIONAL SEAS CONVENTIONS

UNEP's Regional Seas Conventions is one of the most comprehensive action plans for protecting coastal and marine environments. Launched in 1974, these Action Plans cover 18 regions of the world including Arctic, Antarctic, Black Sea, Baltic, Caspian, East Asian Seas, Eastern Africa, Mediterranean, North-East Atlantic, North-East Pacific, Northwest Pacific, Pacific, Red Sea and Gulf of Aden, Persian Gulf, South Asian Seas, South-East Pacific, Western Africa and Wider Caribbean. The first Regional Sea Convention was the Barcelona Convention,1976, an essential part of the Mediterranean Action Plan of 1975. This Convention established the framework model for environmental treaties that underpins other regional sea conventions and is present in several global environmental conventions. All Action Plans are not fully administered by UNEP, but all engaged neighboring nations are aimed at protecting the regional marine environment through a "shared seas" approach in extensive and specific activities. Actions are comprised of multi-sector approaches to both coastal and marine areas, spotlighting the existing identified environmental challenges (UNEP, 2018).

2.2.9   BALTIC MARINE ENVIRONMENT PROTECTION COMMISSION (HELSINKI COMMISSION – HELCOM)

HELCOM, the Helsinki Commission, is the short name for the Baltic Marine Environment Protection Commission, the governing body of the Convention on the Protection of the Marine Environment of the Baltic Sea Area. HELCOM was signed in 1974 to address the increasing environmental challenges stemming from human activities, particularly industrialization. It includes ten members: the nine Baltic Sea countries (Denmark, Estonia, Finland, Germany, Latvia, Lithuania, Poland, Russia and Sweden) and the European Union. Updated in 1992, HELCOM entered into force in 2000 with the declared goals of preventing and eliminating pollution, thus paving the way to the complete ecological restoration of the Baltic Sea. The Convention also applies the polluter-pays principle and promotes the Best Available Technology and Best Environmental Practices. Additionally, its text underlines that the implementation of HELCOM should not cause transboundary pollution outside the Baltic Sea Area (Fullerton, 2014). Under HELCOM's agenda, several guidelines have been developed and made publicly available on a wide range of topics, from reporting of waterborne pollution to determination of "heavy metals" in sediments and or even on the monitoring of radioactive or reprotoxic substances (Pilgrim, 2015).

2.2.10   WTO Notifications under the TBT Agreement for Prevention of Microplastic Pollution

The WTO (World Trade Organization) Agreement on Technical Barriers to Trade (the "TBT Agreement") establishes rules and procedures regarding the development, adoption and application of voluntary product standards, mandatory technical regulations and the procedures (such as testing or certification) used to determine whether a particular product meets such standards or regulations. The aim of the TBT Agreement is to prevent the use of technical requirements as unnecessary barriers to trade. Although the TBT Agreement applies to a broad range of industrial and agricultural products, sanitary and phytosanitary measures and specifications for government procurement are covered under separate agreements. The TBT Agreement rules help to distinguish legitimate standards and technical regulations from protectionist measures. Standards, technical regulations and conformity assessment procedures are to be developed and applied on a nondiscriminatory basis, developed and applied transparently, and should be based on relevant international standards and guidelines, when appropriate (Larsen and Venkova, 2014; EC, 2013).

Under the abovementioned context, the first national regulation on microplastic pollution was adopted in 2015 in the US. Following legislative initiatives in several states, The US Microbead-free Waters Act of 2015 prohibits the manufacture and sale of rinse-off cosmetic products that contain microbeads (Xanthos and Walker, 2017; McDevitt et al., 2017). A microbead is defined as any solid plastic particle that is less than 5 mm in size and is intended to be used to exfoliate or cleanse the human body or any part thereof (Kentin and Kaarto, 2018). The Act also provides that further regulation of microbeads in rinse-off cosmetic products by federal states is not permitted and should be revoked if already in place. In this regulation, the distinction between leave-on and rinse-off products was introduced, proposing that primarily rinse-off products would lead to disposal in waterways.[1]

In the abovementioned scenario, South Korea was the first country to notify the WTO of its proposed prohibition of microbeads in cosmetic products. In the notification database on technical barriers, South Korea announced a ban on microbeads in rinse-off products in October 2016 and in toothpaste in February 2017

---

[1] https://www.congress.gov/congressional-report/114th-congress/house-report/371/1).

(G/TBT/N/KOR/672, 2016; G/TBT/N/KOR/706, 2017). Taiwan followed with notification for new legislation with a reference to the US Microbead-free Waters Act, using more or less the same definitions (G/TBT/N/KOR/672, 2016).

Canada notified the WTO regarding the proposed Microbeads in Toiletries Regulations covering products for cleansing or hygiene and defines microbeads as plastic microbeads that are ≤5 mm in size.[2] Different forms of particles are included, such as solid, hollow, amorphous and solubilized, as well as different functions. Microbeads are distinguished from secondary microplastics, as being manufactured for a specific purpose and application. This definition diverges from commonly used definitions, which describe microbeads often as solid particles with the function of exfoliating and cleansing. The Canadian ban puts microbeads on the list of toxic substances of the Canadian Environmental Protection Act, 1999 (Canadian Environmental Protection Act, 1999). In general, the substances of the following characteristics are considered toxic:

1. Substances may have an immediate or long-term harmful effect on the environment or its biological diversity.
2. Substances may constitute a danger to the environment on which life depends.
3. Substances may constitute a danger in Canada to human life or health.

France was the first EU Member State to adopt a ban, which banned the sale of rinse-off cosmetic products for cleaning or exfoliation that contain solid plastic particles.[3] The ban excludes particles from a natural origin, providing that they are not persistent and do not affect the food chain. The French ban does not specify the size of the particles resulting in all solid plastic particles being banned, also those larger than 5 mm. The ban was notified both to the commission, under the 2015/1535 notification procedure[4], and to the WTO (G/TBT/N/FRA/170, 2016).

Sweden has also announced a ban prohibiting rinse-off cosmetic products containing plastic particles added for exfoliating, cleaning and polishing purposes (G/TBT/N/SWE/132, 2017). Plastic particles are defined as solid particles of plastic 5 mm or less in size and insoluble in water.[5] The Swedish notification refers explicitly to the US and French regulations, and it seems that Sweden has attempted to follow the definition in these regulations.

New Zealand notified the WTO in March 2017 of its proposed ban on microbeads in "wash-down" cosmetic products (G/TBT/N/NZL/77, 2017). In October 2017, New Zealand announced that the proposed ban would be extended to include cleaning products, such as household, car and industrial cleaning products (G/TBT/N/NZL/77/Add.1, 2017). A microbead is defined as a water-insoluble plastic particle less than 5 mm at its widest point,[6] thereby tying in with the regulation in

---

[2] http://laws-lois.justice.gc.ca/eng/regulations/SOR-2017-111/index.html.
[3] http://www.assemblee-nationale.fr/14/dossiers/biodiversite.asp.
[4] https://ec.europa.eu/growth/single-market/barriers-to-trade/tris_en.
[5] https://www.riksdagen.se/sv/dokument-lagar/dokument/svensk%20forfattningssamling/forordning-1998944-om-forbud-mm-i-vissa-fall_sfs-1998-944.
[6] https://www.legislation.govt.nz/regulation/public/2017/0291/latest/whole.html#DLM7490715.

the United States and Canada (Ministry for the Environment, 2017). The United Kingdom has announced legislative proposals for England, Wales, Scotland and Northern Ireland. The Environmental Protection (Microbeads) Regulations 2017 are proposed under the Environmental Protection Act 1990 and follow the US regulation regarding the definition of microplastics and the category of products. Belgium has notified an agreement as a voluntary sector effort to phase out microplastics, initially from rinse-off cosmetic products and gradually from cleaning and maintenance products (TRIS, 2017).

The latest notification came from Italy, proposing to phase out microplastics in exfoliating rinse-off cosmetic products and detergents by January 2020 (G/TBT/N/ITA/33, 2018). Table 2.7 indicates the regulations on microplastics that were either notified or in force.

2.2.11 Convention on the Prevention of Marine Pollution by Dumping of Wastes and Other Matter

The "Convention on the Prevention of Marine Pollution by Dumping of Wastes and Other Matter 1972", the "London Convention" for short, is one of the first global conventions to protect the marine environment from human activities and has been in force since 1975.

In 1996, the "London Protocol" was agreed to further modernize the Convention and, eventually, replace it. Under the Protocol, all dumping is prohibited, except for possibly acceptable wastes on the so-called "reverse list". The Protocol entered into force on 24 March 2006 and there are currently 53 Parties to the Protocol. Its objective is to promote the effective control of all sources of marine pollution and to take all practicable steps to prevent pollution of the sea by dumping of wastes and other matter. Currently, 87 States are Parties to this Convention.[7]

2.2.12 Global Partnership of Marine Litter

GPML, The Global Partnership on Marine Litter, was launched at the United Nations Conference on Sustainable Development in June 2012 in response to a request set out in the Manila Declaration on furthering the implementation of the global program of action for the protection of the marine environment from land-based activities (now called Pollution Free Ecosystems Unit). The partnership is led by a steering committee and UNEP provides secretariat services. The aim of GPML is to protect the global marine environment, human well-being, and animal welfare by addressing the global problem of marine litter and plastic pollution. The key points of GPML are as follows.[8]

1. Providing a platform for cooperation and coordination by sharing ideas, knowledge and experiences and by identifying gaps and emerging issues.
2. Harnessing the expertise, resources and enthusiasm of all stakeholders.

---

[7] https://www.imo.org/en/OurWork/Environment/Pages/London-Convention-Protocol.aspx.
[8] https://www.gpmarinelitter.org/who-we-are.

TABLE 2.7

Regulation on Microplastics (Inforce or Notified)

Country	WTO Notification	Product Category	Definition of Microplastics	References
United States – Microbead-free Waters Act of 2015 (in force)	–	Rinse-off cosmetic products	Microbead: any solid plastic particle that is less than 5 mm in size and is intended to be used to exfoliate or cleanse the human body or any part thereof	McDevitt et al. (2017)
South Korea – Proposed amendments to the "Regulation on Safety Standards of Cosmetics"	G/TBT/N/KOR/672 G/TBT/N/KOR/706	Cleansing products, dental cleansing products	Microbeads: less than or equal to 5 mm in size	G/TBT/N/KOR/672 (2016); G/TBT/N/KOR/706 (2017)
Taiwan – Restrictions on the Manufacture, Import and Sale of Personal Care and Cosmetics Products Containing Plastic Microbeads (in force)	G/TBT/N/TPKM/249	Cosmetics used for washing hair, bathing, face-washing and soap; toothpaste	Microbeads: solid plastic particles used for exfoliation or cleaning of the body wherein the scope of particles' diameter is smaller than 5 mm	G/TBT/N/TPKM/249 (2016)
Canada – Microbeads in Toiletries Regulations (in force)	G/TBT/N/CAN/501	Toiletries, meaning any personal hair, skin, teeth or mouth care products for cleansing or hygiene, including exfoliants	Microbead: plastic microbeads that are ≤5 mm in size, any plastic particle, including different forms such as solid, hollow, amorphous and solubilized	G/TBT/N/CAN/501 (2017) entry into force
France – Decree prohibiting the placing on the market of rinse-off cosmetic products for exfoliation or cleansing that contain solid plastic particles (in force)	G/TBT/N/FRA/170 (2016/543/F-EU notification)	Rinse-off cosmetic products for exfoliation or cleansing	Solid plastic particles, with the exception of particles of natural origin not liable to persist in, or release active chemical or biological ingredients into the environment or to affect animal food chains	G/TBT/N/FRA/170 (2016); Kentin and Kaarto (2018)
New Zealand – Waste Minimization (Microbeads) Regulations 2017	G/TBT/N/NZL/77	Wash-down cosmetic products; cleaning products	Microbead: a water-insoluble plastic particle that is less than 5 mm at its widest point	G/TBT/N/NZL/77 (2017)

(Continued)

TABLE 2.7 (Continued)
Regulation on Microplastics (Inforce or Notified)

Country	WTO Notification	Product Category	Definition of Microplastics	References
Sweden – Draft Regulation prohibiting the placing on the market of rinse-off cosmetics that contain solid plastic particles which have been added for exfoliating, cleaning or polishing purposes	G/TBT/N/SWE/132 (2017/284/S-EU notification)	Rinse-off cosmetic products	Solid particles of plastic which are 5 mm or less in size in any dimension and which are insoluble in water	G/TBT/N/SWE/132, (2017); Kentin and Kaarto (2018)
United Kingdom – The Environmental Protection (Microbeads) Regulations 2017/2018 (England, Wales, Scotland, Northern Ireland)	G/TBT/GBR/28 (2017/353/UK) G/TBT/GBR/29 (2018/42/UK) G/TBT/GBR/30 (2018/48/UK) G/TBT/GBR/32 (2018/208/UK)	Rinse-off personal care products	Microbead: any water-insoluble solid plastic particle of less than or equal to 5 mm in any dimension	Kentin and Kaarto (2018)
Belgium – Draft Sector Agreement to support the replacement of microplastics in consumer products	– (2017/465/B-EU notification)	Not settled		Kentin and Kaarto (2018)
Italy – Draft technical regulation banning the marketing of non-biodegradable and non-compostable cotton buds and exfoliating rinse-off cosmetic products or detergents containing microplastics	G/TBT/N/ITA/33 (2018/258/I-EU notification)	Exfoliating rinse-off cosmetic products and detergents	Water-insoluble solid plastic particles of 5 mm or less, referring to definition in Commission Decision (EU) 2017/1217 of 23 June 2017	G/TBT/N/ITA/33 (2018); Kentin and Kaarto (2018)

3. Making a significant contribution to the achievement of the 2030 Agenda, in particular SDG 14.1 that is by 2025, prevent and significantly reduce marine pollution of all kinds, particularly from land-based activities, including marine debris and nutrient pollution.

2.2.13 CODE OF CONDUCT FOR RESPONSIBLE FISHERIES

This voluntary code of conduct was developed by the UN's Food and Agriculture Organization and was adopted in 1995. The objectives of this code are as follows:

1. Establish principles, following the relevant rules of international law, for responsible fishing and fisheries activities, taking into account all their relevant biological, technological, economic, social, environmental and commercial aspects.
2. Establish principles and criteria for the elaboration and implementation of national policies for responsible conservation of fisheries resources and fisheries management and development;
3. Serve as an instrument of reference to help States to establish or to improve the legal and institutional framework required for the exercise of responsible fisheries and in the formulation and implementation of appropriate measures;
4. Provide guidance which may be used where appropriate in the formulation and implementation of international agreements and other legal instruments, both binding and voluntary;
5. Facilitate and promote technical, financial and other cooperation in the conservation of fisheries resources and fisheries management and development;
6. Promote the contribution of fisheries to food security and food quality, giving priority to the nutritional needs of local communities;
7. Promote the protection of living aquatic resources and their environments and coastal areas;
8. Promote the trade of fish and fishery products in conformity with relevant international rules and avoid the use of measures that constitute hidden barriers to such trade;
9. Promote research on fisheries as well as on associated ecosystems and relevant environmental factors; and
10. Provide standards of conduct for all persons involved in the fisheries sector.

However, owing to the fact that the Food and Agriculture Organization has no legislative authority and because of the voluntary nature of the Code, even those who have publicly embraced this Code cannot be forced to implement measures to successfully achieve any of the detailed objectives (Gall and Thompson)[9, 10].

---

[9] https://www.europarl.europa.eu/RegData/etudes/STUD/2020/658279/IPOL_STU(2020)658279_EN.pdf.
[10] https://www.fao.org/3/v9878e/v9878e00.htm#2.

2.2.14 Convention on Biological Diversity

The Convention on Biological Diversity entered into force in 1993 and has been ratified by 196 nations. Its main objectives include:

1. The conservation of biological diversity
2. The sustainable use of the components of biological diversity
3. The fair and equitable sharing of the benefits arising out of the utilization of genetic resources.

In 2016 the Conference of Parties urged members to implement within their national jurisdictions measures to prevent and mitigate the impacts of marine debris on marine and coastal biodiversity[11].

2.2.15 Strategic Approach to International Chemicals Management (SAICM)

SAICM, Strategic Approach to International Chemicals Management, is a policy framework to promote chemical safety around the world. It was by the First International Conference on Chemicals Management on 6 February 2006 in Dubai. SAICM was developed by a multi-stakeholder and multi-sectoral Preparatory Committee and supports the achievement of the 2020 goal agreed at the 2002 Johannesburg World Summit on Sustainable Development. SAICM's overall objective is the achievement of the sound management of chemicals throughout their life cycle so that by the year 2020, chemicals are produced and used in ways that minimize significant adverse impacts on the environment and human health. SAICM comprises the Dubai Declaration on International Chemicals Management, expressing high-level political commitment to SAICM, and an overarching Policy Strategy which sets out its scope, needs, objectives, financial considerations underlying principles and approaches, and implementation and review arrangements. Objectives of SAICM are grouped into the following themes:

1. Risk reduction
2. Knowledge and Information
3. Governance
4. Capacity-building and technical cooperation
5. Illegal international traffic

The Declaration and Strategy are accompanied by a Global Plan of Action that serves as a working tool and guidance document to support the implementation of SAICM and other relevant international instruments and initiatives. Activities in the plan are to be implemented, as appropriate, by stakeholders, according to their applicability.[12]

---

[11] https://www.un.org/en/observances/biological-diversity-day/convention.
[12] https://www.saicm.org/About/Overview/tabid/5522/language/en-US/Default.aspx.

2.2.16 STOCKHOLM CONVENTION

Stockholm Convention on Persistent Organic Pollutants is an international environmental treaty signed in 2001 and entered into force in May 2004. The objective of the Stockholm Convention is to protect human health and the environment from persistent organic pollutants. The main provisions of the convention are as follows[13]:

1. Annex A allows for the registration of specific exemptions for the production or use of listed persistent organic pollutants, in accordance with that Annex and Article 4, bearing in mind that special rules apply to polychlorinated biphenyls. The import and export of chemicals listed in Annex A can occur under specific restrictive conditions, as set out in paragraph 2 of Article 3.
2. The convention prohibits and/or eliminates the production, use, import and export of intentionally produced POPs (Annex A to the Convention (Article 3).
3. Restrict the production, use, import and export of the intentionally produced POPs. (listed in Annex B to the Convention (Article 3). Annex B allows for the registration of acceptable purposes for the production and use of the listed POPs.
4. Reduce or eliminate releases from unintentionally produced POPs. (listed in Annex C to the Convention (Article 5).
5. The Convention promotes the best available techniques and best environmental practices for preventing releases of POPs into the environment.
6. Ensure that stockpiles and wastes containing or contaminated with POPs are managed safely and in an environmentally sound manner (Article 6).
7. The Convention requires that such stockpiles and wastes be identified and managed to reduce or eliminate POPs releases from these sources. The Convention also requires that wastes containing POPs are transported across international boundaries considering relevant international rules, standards and guidelines.
8. To target additional POPs (Article 8).
9. The Convention provides detailed procedures for listing new POPs in Annexes A, B and/or C. A committee composed of experts in chemical assessment or management – the Persistent Organic Pollutants Review Committee, is established to examine proposals for the listing of chemicals, in accordance with the process set out in Article 8 and the information requirements specified in Annexes D, E and F of the Convention.
10. Other provisions of the Convention relating to the development of implementation plans (Article 7), information exchange (Article 9), public information, awareness and education (Article 10), research, development and monitoring (Article 11), technical assistance (Article 12), financial resources and mechanisms (Article 13), reporting (Article 15), effectiveness evaluation (Article 16) and non-compliance (Article 17).

---

[13] http://www.pops.int/TheConvention/Overview/tabid/3351/Default.aspx.

2.3 POLICY CONSIDERATIONS AND RECOMMENDATIONS

The challenge of plastic production control and its environmental impact requires immediate attention. Besides being level of effort, all individual and collective efforts that have been done focus on prevention, debris monitoring and cleanup initiatives, mitigation and educational awareness (Bergmann et al., 2015).

But efforts should also make further with the possibility of science-based or science-informed solutions. Initiatives should be carried out from a holistic perspective as follows[14]:

1. Stakeholder engagement should be ensured by calls for awareness campaigns and policy discussions.
2. Underpin the exact causes of plastic waste and solve them one by one.
3. Reduction targets should be established to control the use of plastics.
4. Stricter measures should be established to combat single-dose packaging, focusing on the food chain supply and cosmetic industry.
5. The right of the customer to return the plastic packaging to the retailer should be secured by policy implementation.
6. Comprehensive and transparent regulations should be made available for the labeling of bio-based and biodegradable plastics. The public should be made aware of it by improving their educational level.
7. Zero-waste favoring Activities should be developed that stimulate zero-waste production. This may be achieved by harmonizing the regulatory framework schemes and promoting the 4R rule.
8. Revenues from fines and levies should be made available for activities associated with zero plastic waste like awareness programs, financing the recycling industry or supporting specific environmental projects. The economic preference should be provided for the use of virgin polymers by contemplating the application of progressive taxes on these materials, thus reducing the industry's impulses on the unhindered use of plastics in manufacture and packaging.
9. Should promote fiscal incentives or tax rebates for manufacturers, suppliers and retailers who develop and implement zero-waste transition activities
10. Should have some flexible regulations that can monitor the progress and effectiveness of implemented policies and amend or adjust them where needed.
11. Should invest funds in establishing modern infrastructure that can facilitate the collection, separation and processing of plastic waste from neglected rural areas.

ACKNOWLEDGMENTS

This work was supported by a grant from the National Research Foundation of Korea (NRF) grant funded by the Korea government (MSIT) (No. NRF-2020R1A2C1013851).

---

[14] https://www.europarl.europa.eu/RegData/etudes/STUD/2020/658279/IPOL_STU(2020)658279_EN.pdf.

REFERENCES

Agence France-Presse, 2017. Sri Lanka bans plastic after garbage crisis. *Agence France-Presse*. http://www.dailymail.co.uk/wires/afp/article-4843304/SriLanka-bans-plastic-garbage-crisis.html.

Alexander, P., 2017. New York City tries to ban Styrofoam: It's Déjà Vu all over again. *Huffington Post*, 25 October. https://www. huffingtonpost.com/entry/new-york-city-triesto-ban-styrofoam-its-d%C3%A9j%C3%A0-vu_us_59f08cefe4b02ace788ca8ea.

Alicia, 2011. Plastics forever. Road to Ethiopia (blog). Available from: https://ethiopia.limbo13.com/index.php/plastics_are_forever/.

Alpagro Plastics, 2016. Plastic bags soon to be banned? Available from: https://www.alpagro-plastics.be/en/news/carrier-bags/plastic-bags-soon-to-be-banned/90.

American Samoa Environmental Protection Agency, 2011. http://www.epa.as.gov/press-releases-2011.

Antigua Nice Ltd, 2017. Stages and implementation of Styrofoam ban. Available from: http://www.antiguanice.com/v2/client.php?id=806&news=10298.

Associated Press, 2015. Kicking the plastic: Senegal among latest to ban flimsy bags, 2015. Available from: http://www.dailymail.co.uk/wires/ap/article-3102733/Kicking-plastic-Senegal-latest-ban-flimsy-bags.html.

Babin, J., 2017. New York City reinstates Styrofoam ban. *WNYC News*, 13 May. http://www.wnyc.org/story/new-york-city-reinstates-styrofoamban/.

Bari, P., 2018. Plastic ban has worked in Sikkim but not in Delhi, finds Pune-based NGO. *Hindustan Times*. Available from: https://www.hindustantimes.com/pune-news/plastic-ban-has-worked-in-sikkim-but-not-in-delhi-finds-pune-based-ngo/storyEGV9D4hl1yhUFFLGt9vcTK.html.

BBC News, 2013. Mauritania bans plastic bag use. Available from: http://www.bbc.com/news/world-africa-20891539.

BBC News, 2015. Plastic bag charge in Scotland sees usage cut by 80%, 2015. Available from: http://www.bbc.com/news/uk-scotland-34575364.

Belize Press Office, 2018. Joint press release: Phasing out of single-use plastic bags and Styrofoam and plastic food utensils. Available from: https://www.facebook.com/GOBPressOffice/photos/a.150654578303387.21501.149350998433745/1627401217295375/?type=3&theater.

Bergmann, M., Gutow, L., Klages, M., 2015. *Marine Anthropogenic Litter*. Springer Nature.

Black, E., 2016. Indonesia's plastic bag tax not enough, say experts. *Southeast Asia Globe*. Available from: https://www.theguardian.com/ environment/the-coral-triangle/2017/mar/02/indonesia-pledges-us1-billion-a-year-to-curb-ocean-waste.

Boisvert, M., 2014. d'Ivoire chokes on its plastic shopping bags. InterPress Service. Available from: http://www.ipsnews.net/2014/09/cote-divoire-chokes-on-its-plastic-shopping-bags/.

Braun, Y. A., Traore, A. S., 2015. Plastic bags, pollution, and identity: Women and the gendering of globalization and environmental responsibility in Mali. *Gender & Society*, 29(6), 863–887.

Brizga, J., 2016. *Packaging Tax in Latvia*. Institute for European Environmental Policy publication.

Bulgaria's Environment Ministry Reports Substantial Reduction in Plastic Bag Use, 2015. Novinite. Available from: http://www.novinite.com/articles/168268/Bulgaria's+Environment+Ministry+Reports+Substantial+Reduction+in+Plastic+Bag+Use.

Canadian Environmental Protection Act, 1999. https://laws-lois.justice.gc.ca/eng/acts/c-15.31/.

Cann, G., 2017. Nearly half the country mayors join call for compulsory charge on plastic bags. *Stuff*. Available from: https://www.stuff.co.nz/environment/94015572/nearly-half-the-countrys-mayors-join-call-for-compulsory-charge-on-plastic-bags.

Carreon, B. H., 2017. Palau bans use of plastic bags. *Marianas Business Journal*. https://mbjguam.com/2017/11/27/palau-bans-use-of-plastic-bags/.

Cherone, H., Wetli, P., 2017. Chicago's plastic bag tax is working – Big time, study shows. *DNA info*, 24 April. https://www.dnainfo.com/chicago/20170424/lincoln-square/were-using-42-percent-fewer-bags-since-7-cent-tax-started-citystudy-says.

Chitotombe, J. W., Gukurume, S., 2014. The plastic bag 'ban'controversy in Zimbabwe: An analysis of policy issues and local responses. *International Journal of Development and Sustainability*, 3(5), 1000–1012.

Chiyal, B. L. S., 2017. To help conserve Lake Town bans plastic bags. *Global Press Journal*. https://globalpressjournal.com/americas/guatemala/to-help-conserve-lake-atitlan-town-bans-plastic-bags/.

Chow, L., 2017. Chile bans plastic bags in 100+ areas. *EcoWatch*. Available from: https://www.ecowatch.com/chile-plastic-ban-2501794220.html.

Clavel, 2014. Think you can't live without plastic bags? Consider this: Rwanda did it. *The Guardian*. Available from: https://www.theguardian.com/commentisfree/2014/feb/15/rwanda-banned-plastic-bags-so-can-we.

Clayton, R., 2017. Countdown to ban all single use plastic bags by 2018. *Stuff*. Available from: https://www.stuff.co.nz/business/97528655/countdown-to-ban-all-single-use-platic-bags-by-2018.

Clean Bhutan, 2015. Ban on plastic 1999. *Clean Bhutan*, 21 April. http://www.cleanbhutan.org/ban-on-plastic-1999/.

CNA News Service, 2018. Bill on plastic bags was rushed through without enough debate, Theopemptou says. *CyprusMailOnline*. Available from: http://cyprus-mail.com/2018/01/23/bill-plastic-bags-rushed-without-enough-debate-theopemptou-says/.

Coker, O., 2017. NEA: plastic bags still banned. *The Standard*. Available from: http://standard.gm/site/2017/11/16/nea-plastic-bags-still-banned/ [Accessed 16 November 2017].

Colbert, G., 2016. Cameroon: Bagging it after the plastic ban. *African Argument*. Available from: http://africanarguments.org/2016/11/30/cameroon-bagging-it-after-the-plastic-ban/.

Cooper, L., 2017a. Victoria jus announced it will ban single-use plastic bags. *HuffPost Australia*. Available from: http://www.huffingtonpost.com.au/2017/10/17/victoria-just-announced-it-will-ban-single-use-plastic-bags_a_23245619/.

Cooper, L., 2017b. Western Australia will ban single-use plastic bags from next year. *HuffPost Australia*. Available from: http://www.huffingtonpost.com.au/2017/09/12/western-australia-will-ban-single-use-plastic-bags-from-next-year_a_23205446/.

COSTA, 2015. The impact of debris on marine life. *Marine Pollution Bulletin*, 92(1–2), 170–179.

da Costa, J., 2017. Microplastics–occurrence, fate and behaviour in the environment. In *Comprehensive Analytical Chemistry* (Eds APT Rocha-Santos, AC Duarte). Elsevier, Amsterdam. pp. 1–24.

Deccan, C., 2016. Ban on plastic bags from today in Kochi. *Deccan Chronicle*. Available from: http://www.deccanchronicle.com/nation/in-other- news/011016/ban-on-plastic-bags-from-today-in-kochi.html.

Deepika, K. C., 2017. Plastic 'garbage bags' are available in plenty online. *The Hindu*. Available from: http://www.thehindu.com/news/national/karnataka/plastic-garbage-bags-are-available-in-plenty-online/article19338420.ece.

DHNS New Delhi, 2017. Supreme Court rejects plea against Karnataka decision to ban plastic. *DeccanHerald*. Available from: http://www.deccanherald.com/content/641501/sc-rejects-plea-against-karnataka.html.

Dikgang, J., Leiman, A., Visser, M., 2012. Analysis of the plastic-bag levy in South Africa. *Resources, Conservation and Recycling*, 66, 59–65.

Dikgang, J., Martine, V., 2010. Behavioral response to plastic bag legislation in Botswana. Environment for Development. Discussion Paper Series. Available from: http://www.rff.org/files/sharepoint/WorkImages/Download/EfD-DP-10-13.pdf.

Duboise, T., 2010. Thompson, subarctic city in Canada, bans plastic bags. *Friends Across The Globe*, 28 December. http://plasticbagbanreport.com/thompson-subartic-city-in-canada-bansplastic-bags/.

Duboise, T., 2012. Leaf rapids, Manitoba plastic bag ban. *Friends Across The Globe*, 11 June. http://plasticbagbanreport.com/leaf-rapids-manitobaplastic-bag-ban/.

Eastaugh, S., 2016. France becomes firs country to ban plastic cups and plates. *CNN News*. Available from: http://edition.cnn.com/2016/09/19/europe/france-bans-plastic-cups-plates/index.html.

Ellis, M., 2016. Morocco Bans Plastic Bags to Combat Plastic Waste Pollution. Better

EnviroNews Nigeria, 2017. Cape Verde outlaws plastic bags, Congo launches 'Blue Fund'. environewsnigeria.com, 6 April. http://www.environewsnigeria.com/cape-verde-outlaws-plasticbags-congo-launches-blue-fund/; http://www.govmu.org/English/News/Pages/Mauritius-bans-the-use-of-plastic-bags.aspx.

Ethiopian News Agency, 2016. Over dozen plastic factories producing sub-standard bags in AddisAbaba. *Ethiopian News Agency*. Available from: http://www.ena.gov.et/en/index. php/environment/item/1058-over-dozen-plastic-factories-producing-sub-standard-bags-in-addis-ababa.

Euronews, 2016. France bans plastic bags, what about the rest of the EU? *Euronews*. Available from: http://www.euronews.com/2016/06/30/france-bans-plastic-bags-what-about-the-rest-of-the-eu.

European Commission (EC), 2013. Commission Staff Working Document Impact Assessment for a Proposal for a Directive of the European Parliament and of the Council amending Directive 94/62/EC on packaging and packaging waste to reduce the consumption of lightweight plastic carrier bags. *EUR-Lex*. http://eur-lex.europa.eu/legal-content/EN/ALL/?uri=CELEX:52013SC0444.

Express Tribune, 2018. Sind government imposes ban on use of plastic bags. *The Express Tribune*. Available from: https://tribune.com.pk/story/1666275/1-sindh-govt-imposes-ban-use-plastic-bags/.

Fickling, D., 2003. Tasmania carries the eco- fight by banning plastic bags carries hopes of environment. *The Guardian*. Available from: https://www.theguardian.com/world/2003/apr/29/australia.Davidficklin [Accessed 29 April 2003].

Fikrejesus, 2017. Banning plastic bags. The case of Eritrea. *Fikre*. Available from: https://fiqre4eriwordpress.com/2017/08/29/banning-plastic-bags-the-case-of-eritrea/.

Fullerton, K., 2014. Reflecting on Rwanda's plastic bags ban. *International Development Journal*. https://idjournal.co.uk/2017/04/24/reflectingrwandas-plastic-bags-ban/.

Fundira, M., 2016. Montreal bylaw on plastic bag ban passed. *CBC News*, 24 August. http://www.cbc.ca/news/canada/montreal/montreal-plasticbag-ban-1.3733504.

G/TBT/N/CAN/501, 2017. Committee on Technical Barriers to Trade.

G/TBT/N/FRA/170, 2016. Committee on Technical Barriers to Trade. http://tbtims.wto.org/en/RegularNotifications/View/92287.

G/TBT/N/ITA/33, 2018. Committee on Technical Barriers to Trade, 2018. http://tbtims.wto.org/en/RegularNotifications/View/142886?FromAllNotifications=True.

G/TBT/N/KOR/672, 2016. Committee on Technical Barriers to Trade. https://members.wto.org/crnattachments/2016/TBT/KOR/16_4194_00_x.pdf.

G/TBT/N/KOR/706, 2017. Committee on Technical Barriers to Trade. https://members.wto.org/crnattachments/2017/TBT/KOR/17_0627_00_x.pdf.

G/TBT/N/NZL/77, 2017. Committee on Technical Barriers to Trade. http://tbtims.wto.org/en/RegularNotifications/View/98853?FromAllNotifications=True.

G/TBT/N/NZL/77/Add.1, 2017. Committee on Technical Barriers to Trade. http://tbtims.wto.org/en/ModificationNotifications/View/138388?FromAllNotifications=True.

G/TBT/N/SWE/132, 2017. Committee on Technical Barriers to Trade; Technical Regulation Information System, 'Notification Number 2017/284/S. http://tbtims.wto. org/en/ModificationNotifications/View/137307.

G/TBT/N/TPKM/249, 2016. Committee on Technical Barriers to Trade, Attachment for English text of legislation. Available from: https://members.wto.org/crnattachments/ 2016/TBT/TPKM/16_4322_00_e.pdf.

Gerrard, J., 2018. Spain's plastic bag ban postponed amid lack of support. *EuroWeek*. Available from: https://www.euroweeklynews.com/news/on-euro-weekly-news/spain-news-in-english/1471777-spain-s-plastic-bag-ban-postponed-amid-lack-of-support.

Guinea-Bissau: Retailers hit back at plastic bag ban, 2017. TrendType. Available from: https:// trendtype.com/2017/04/17/guinea-bissau-retailers-hit-back-at-plastic-bag-ban/.

Harford, S., 2017. Coop Sweden works to reduce plastic bag usage. *European Supermarket Magazine*. https://www.esmmagazine.com/coop-sweden-works-reduce-plastic-bags/43980.

Harvey, C., 2017. The US state that banned banning plastic bags. *Independent*, 2 January. http://www.independent.co.uk/news/world/americas/ michigan-the-us-state-that-just-banned-banningplastic-bags-a7505611.html.

Hasan, Y. M., 2017. Somaliland: Despite ban use of plastic bag continues unabated. *Somaliland Sun*. Available from: http://www.somalilandsun.com/2017/02/06/somaliland-despite-ban-use-of-plastic-bag-continues-unabated/.

Haskell, P., 2014. Galapagos bans plastic bags. Galapagos Conservation Trust. Available from: http://galapagosconservation.org.uk/galapagos-bans- plastic-bags/.

Hayne, J., 2017. What difference did the plastic bag ban make to Canberra's waste? *ABC News (Australia)*. Available from: http://www.abc.net.au/news/specials/curious-canberra/2017-04-10/what-difference-did-the-plastic-bag-ban-make-to-canberras-waste/8392804.

Hong, T. H., 2016. Bandung bans Styrofoam, Pop Mie to switch to paper packaging. *Minime Insights*. Available from: http://edition.cnn.com/2017/08/16/asia/melati-isabel-wijsen-bali/index.html.

Huffadine, L., 2017. New World supermarket asks for feedback on plastic bags. *Stuff*. Available from: https://www.stuff.co.nz/business/industries/96625069/new-world-supermarket-asks-for-feedback-on-plastic-bagsx [Accessed 14 September 2017].

India, Ministry of Environment, Forest and Climate Change, 2016. Notification on Plastic Waste Management Rules, published in the Gazette of India, Part-II, Section-3, Sub-section (i), March 18. http://www.indiaenvironmentportal.org.in/files/file/ Plastic%20Waste%20Management%20Rules%20 2016.pdf.

IRIN, 2010. Just say no to plastic bags. *Irin News*. Available from: http://www.irinnews.org/ news/2010/11/24/just-say-no-to-plastic-bags [Accessed 22 August 2021].

IRIN, 2011. Plastics proliferate despite ban. *Irin News*, 2 March. http://www.irinnews. org/report/92072/bangladesh-plastics-proliferatedespite-ban

Jambeck, J. R., et al., 2015. Plastic waste inputs from land into the ocean. *Science*, 347(6223), 768–771.

Jayasekara, S., 2017. Use of polythene, rigifoam, shopping bags banned. *Daily Mirror*. Available from: http://www.dailymirror.lk/132675/Use-of-polythene-rigifoam-shopping-bags-banned. India, Ministry of Environment, Forest and Climate [Accessed 25 August 2021].

Jayasiri, H. B., Purushothaman, C. S., Vennila, A., 2013. Plastic litter accumulation on high-water strandline of urban beaches in Mumbai, India. *Environmental Monitoring and Assessment*, 185(9), 7709–7719.

Jong, H. N., 2017. Government looks to expand plastic bag ban nationwide. *The JakartaPost*. Available from: http://www.thejakartapost.com/news/2017/02/20/govt-looks-expand-plastic-bag-ban-nationwide.html.

Kao, E., 2013. China to lift ban on disposable Styrofoam lunch boxes. *South China Morning Post*, 14 March. http://www.scmp.com/news/china/article/1190498/china-lift-ban-disposablestyrofoam-lunch-boxes.

Karuhanga, J., 2017. Eala votes to ban polythene bags. *The New Times*. Available from: http://www.newtimes.co.rw/section/read/213545/ [Accessed 22August 2021].

Kentin, E., Kaarto, H., 2018. An EU ban on microplastics in cosmetic products and the right to regulate. *Review of European, Comparative & International Environmental Law*, 27(3), 254–266.

Khattak, S., 2017. Khyber-Pakhtunkhwa bans plastic products, again. *The Express Tribune*. Available from: https://tribune.com.pk/story/1441526/k-p-bans-plastic-products/.

Kiprop, V., 2017. Finally, Kenya effects ban on plastic bags. *The East African*. Available from: http://www.theeastafrican.co.ke/business/Kenya-effects-ban-on-plastic-bags-/2560-4086512-10oy0x4/index.Html.

Kis, 2015. Environmental protection product fee on invoices. *Finacont Tax Newsletter*. Available from: http://www.finacont.com/finacont_uploads/files/hirlevel/Tax%20Newsletter%20-%20environmental%20protection%20product%20fee%20on%20invoices%2015062015.pdf.

Lall, R.R., 2013. Haiti police raid warehouses in plastics ban crackdown. *The Guardian*. Available from: https://www.theguardian.com/world/2013/aug/15/haiti-police-raid-plastics-ban-crackdown.

Larsen, J., 2014. Plastic bag bans or fees cover 49 million Americans. Earth Policy Institute, 1 October. http://www.earth-policy.org/data_highlights/2014/highlights49.

Larsen, J., Venkova, S., 2014. The downfall of the plastic bag: A global picture. Earth Policy Institute, 1 May. http://www.earth-policy.org/ plan_b_updates/2013/update123.

LégiBénin, 2018. Vers la répression: A partir du 19 août 2018, emprisonnement ferme et amende pour quiconque détient en sa possession des sachets plastiques! *Journal officiel de la république de Bénin*. Available from: http://www.legibenin.net/index.php/component/k2/item/132-emprisonnement-ferme-et-amende-pour-quiconque-detient-en-sa-possession-des-sachets-plastiquesâ.

Lithuania, Ministry of the Environment, 2016. Retailers will no longer be allowed to hand out plastic carrier bags free of charge. Available from: http://www.am.lt/VI/en/VI/article.php3?article_id=823.

Los Angeles Times Editorial Board, 2017. It's been a year since California banned single-use plastic bags. The world didn't end. *Los Angeles Times*, 16 November. http://www.latimes.com/opinion/editorials/la-ed-plastic-bag-ban-anniversary20171118-story.html.

Lu, H., 2016. Uruguay to ban free plastic bags. *Xinhuanet*, 29 July. http://www.xinhuanet.com/english/photo/2016–07/29/c_135549340.htm.

Mahesh, P. B., et al., 2015. Kolkata and environment, plastic menace. Toxics Link. http://toxicslink.org/docs/Plastic-Report-RevisedJustified-june-2015.pdf.

Malkin, E., 2009. Unveiling a plastic bag ban in Mexico City. *The New York Times*. Available from: https://green.blogs.nytimes.com/2009/08/21/unveiling-a-platic-bag-ban-in-mexico-city/.

Manifava, D., 2017. Plastic bag charge to be introduced from January. Ekathimerini.com. http://www.ekathimerini.com/220859/article/ekathimerini/business/plastic-bag-charge-to-be-introduced-from-january.

Marica, I., 2018. Romania to ban thin plastic bags. Romania-Insider.com. Available from: https://www.romania-insider.com/romania-ban-thin-plastic-bags/.

Martin, G., 2009. Second Argentina province to ban non-biodegradable plastic bags. *ICIS*, 5 November. https://www.icis.com/resources/news/2009/11/05/9261344/second-argentina province-to-ban-non-biodegradable-plastic-bags/.

Martina, K., 2017. Tunisia bans disposable plastic shopping bags. *Treehugger*. Available from: https://www.treehugger.com/environmental-policy/tunisia-bans-disposable-plastic-shopping-bags.html.

Martinho, G., Balaia, N., Pires, A., 2017. The Portuguese plastic carrier bag tax: The effects on consumers' behavior. *Waste Management*, 61, 3–12.

Masai, L. Y., 2015. Somaliland: Ban on plastic bags imposed. *Somaliland Sun*. Available from: http://www.somalilandsun.com/politics1/44-government/government/7236-somaliland-ban-on-plastic-bags-imposed.

Masud, S., 2017. Danger of using plastic bags. *Environment*, 18 December. https://www.sci-remag.com/single-post/2017/12/18/Dangers-of-using-plastic-bags.

McDevitt, J. P., Criddle, C. S., Morse, M., Hale, R. C., Bott, C. B., Rochman, C. M., 2017. Addressing the issue of microplastics in the wake of the microbead-free waters act- a new standard can facilitate improved policy. *Environmental Science & Technology*, 51(12), 6611–6617.

Mhofu, S., 2017. Zimbabwe temporarily lifts ban on foam food containers. *Voa News*. Available from: https://www.voanews.com/a/zimbabwe- environment-styrofoam-eps-containers/3949063.Html.

Ministry for the Environment, 2017. Cabinet Paper: Prohibiting the sale and manufacture of wash off products containing plastic microbeads. Available from: http://www.mfe.govt. nz/node/23631.

Morris, S., 2015. Plastic bag use down 70% in Wales since charges began. *The Guardian*. Available from: https://www.theguardian.com/environment/2015/sep/04/plastic-bag-use-down-70-wales-since-charges-began.

Mozambique News Agency, 2015. Mozambique will limit use of plastic bags. *Asoko Insight*, Available from: https://asokoinsight.com/news/mozambique-will- limit-use-of-plastic-bags.

Naeem, W., 2013. Ban on plastic bags from April 1. *The Express Tribune*, 1 February. https:// tribune.com.pk/story/501287/ban-on-plastic-bagsfrom-april–1/.

Naik, Y., 2017a. India just banned all forms of disposable plastic in its capital. *Mumbai Mirror*, 25 January. Available from: https://mumbaimirror.indiatimes.com/ mumbai/cover-story/state-bans-plastic-bags-from-gudi-padwa/articleshow/60487040.cms.

Naik, Y., 2017b. State bans plastic bags from Gudi Padwa. *Mumbai Mirror*, 13 September. https://mumbaimirror.indiatimes.com/mumbai/cover-story/state-bans-plastic-bags-from-gudi-padwa/articleshow/60487040.cms.

Namara, E., 2016. As Uganda's plastic bag ban takes hold, women create woven, reusable bags. *Global Press Journal*. Available from: https://globalpressjournal.com/africa/ uganda/as-ugandas-plastic-bag-ban-takes-hold-women-create-woven-reusable-bags/.

NDTV India, 2010. Haryana bans use of plastic carry bags. *NDTV*. Available from: https:// www.ndtv.com/india-news/haryana-bans-use-of-plastic-carry-bags-440253.

News Expats, 2017. Czech Republic says goodbye to free plastic bags. https://news.expats. cz/community/czechia-says-goodbye-tofree-plastic-bags/.

Nforngwa, E., 2014. Cameroon grapples with ban on plastic bags. *Al Jazeera*. Available from: http://www.aljazeera.com/news/africa/2014/09/ cameroon-grapples-with-ban-plastics-bags-2014941028535071.html.

Ng, K. L., Obbard, J. P., 2006. Prevalence of microplastics in Singapore's coastal marine environment. *Marine Pollution Bulletin*, 52(7), 761–767.

O'Neill, B., 2016. Economic instruments to reduce usage of plastic bags: The Irish experience. Presentation, informal meeting of EU waste directors, Brussels. Available from: http://ec. europa.eu/transparency/regexpert/index.cfm?do=groupDetail.groupDetailDoc&id= 27612&no=6.

Obateru, T., 2016. Nigeria and the menace of plastic bags. Actions against plastic bags. Available from: https://globalreportingblog.wordpress.com/2016/04/14/nigeria-and-the-menace-of-plastic-bags/.

Panapress, 2014. Niger: Government bans production, import, trade, use of plastic bags. http:// www.panapress.com/Niger--Govt.-bansproduction, -import, -trade, -use-of-plastic-bags--12-630408542-40-lang2-index.html.

People's Daily Online, 2009. Myanmar works for environment conservation. Available from: https://en.people.cn/90001/90777/90851/6830589.html [Accessed 2 December 2009].

Petrone, A., 2015. Brazil launches ban on traditional plastic bags. *The Global Grid*. Available from: http://theglobalgrid.org/city-of-sao-paulo-brazil- launches-ban-on-traditional-plastic-bags/.

Pieters, J., 2017. Dutch ban on free plastic bags sees 71% drop in use. *The Netherlands Times*. https://nltimes.nl/2017/04/18/dutch-banfree-plastic-bags-sees-71-pct-drop-use Published April 18th, 2017 [Online].

Pilgrim, S., 2015. Smugglers work on the dark side of Rwanda's plastic bag ban. *Al Jazeera America*, 25 February. http://america.aljazeera.com/ articles/2016/2/25/rwanda-plastic-bag-ban.html.

Plastic Portal, 2018. We pay for plastic bags in Slovakia and in the Czech Republic. PlasticPortal.eu, 16 January. http://www.plasticportal.eu/en/wepay-for-plastic-bags-in-slovakia-and-in-the-czechrepublic/c/4795/ [Accessed 10 January 2021].

PTI, 2016. NGT for complete ban on plastic bags in Punjab, Haryana. *The Tribune*, 17 January. http://www.tribuneindia.com/news/punjab/ ngt-for-complete-ban-on-plastic-bags-in-punjabharyana/184593.html.

Quillen, S., 2017. Tunisia bans plastic bags in supermarkets. *The Arab Weekly*. Available from: https://thearabweekly.com/tunisia-bans-plastic-bags-supermarkets.

Raz-Chaimovich, M., 2017. 42% of Israelis don't buy plastic bags. *Globes*. Available from: http://www.globes.co.il/en/article-42-of-israelis-dont-buy-plastic-bags-1001184395.

Reuters, 2011. Congo bans plastic bags to fight pollution. http://www.reuters. com/article/ozatp-congo-environment-plastic-20110602- idAFJOE7510G320110602.

Reyes, C., 2017. First city in Mexico to ban use of plastic bags. *Science and Technology*. Available from: https://www.riviera-maya-news.com/queretaro-first-city-in-mexico-to-ban-use-of-plastic-bags/2017.html.

Richards, B., 2017. Concerns about single use bags prompt review into use of thicker bags at retailers. *The Mercury*. Available from: http://www. themercury.com.au/news/tasmania/concerns-about-single-use-bags-prompt-review-into-use-of-thicker-bags-at-retailers/news-story/8439ed6547 2ac35e66de62d57b5ac178.

Rigby, M., Adam, S., Kate, O. T., 2017. War on waste: NT environmental groups claim plastic bag ban has failed. *ABC News (Australia)*. Available from: http://www.abc.net.au/news/2017-05-24/war-on-waste-nt-plastic-bag-ban-fails-say-environment- groups/8553614.

Seattle, Washington USA, 2017. Bag requirements. Available from: http://www.seattle.gov/util/MyServices/Recycling/ReduceReuse/PlasticBagBan/index.htm.

Serju, C., 2017. Government wants public input on plastic ban. *The Gleaner*. Available from: http://jamaica-gleaner.com/article/lead-   stories/20170817/govt-wants-public-input-plastic-ban.

Siqueira, L., 2011. Plastic bag bans in Brazil spark debate. *ICIS News*. Available from: https://www.icis.com/resources/news/2011/07/22/9479546/plastic-bag-bans-in-brazil-spark-debate/.

Sisay, A., 2016. Ethiopia to restrict the use of plastic products. *Africa Review*. Available from: http://www.africareview.com/news/Ethiopia-to-restrict-the-use-of-plastic-products/979180-3137822-jumv3o/index.html.

Smith, J., 2018. Greek shoppers responding to plastic bag ban. GreekReporter.com. Available from: http://greece.greekreporter.com/2018/02/06/greek-shoppers-responding-to-plastic-bag-tax/.

Smithers, R., 2016. England's plastic bag usage drops 85% since 5p charge introduced. *The Guardian*. Available from: https://www.theguardian.com/environment/2016/jul/30/england-plastic-bag-usage-drops-85-per-cent-since-5p-charged-introduced [Accessed 30 July 2016].

SPREP, 2018. Vanuatu plastic bag ban now in effect. Available from: http://www.sprep.org/waste-management-pollution-control/vanuatu-plastic-bag-ban-now-in-effect.

Styrofoam and Plastic Products Prohibition Act, 2016. Marshall Islands. Available from: https://rmiparliament.org/cms/images/LEGISLATION/BILLS/2016/2016–0028/ Styrofoamand Plastic Products ProhibitionAct2016.Pdf.

Styrofoam ban in Sikkim, 2016. The telegraph. Available from: https://www.telegraphindia.com/1160525/jsp/siliguri/story_87417.jsp.

Sun, N. Y., 2015. China's new plastics ban in Jilin province boosts bioplastics sector. *Plasticsnews Europe*. Available from: http://www.plasticsnewseurope.com/article/ 20150114/PNE/301149981/chinas-new-plastics-ban-in-jilin-province-boosts-bioplastics-sector.

Surfrider Foundation Europe, 2017. Enough excuses: Time for Europe to act against plastic bag pollution. International Plastic Free Day 2017. Available from: https://www.surfrider.eu/wpcontent/uploads/2017/06/report_EUMemberStateslegislations_PlasticBags_web_en.pdf.

Tavella, J. M., 2017. BYO bag? Plastic shopping bags banned in Buenos Aires. *The Bubble*. Available from: http://www.thebubble.com/byo-bag-plastic-shopping-bags-banned-in-buenos-aires/.

Technical Regulation Information System (TRIS), 2017. Notification Number 2017/0465/B. Available from: https://ec.europa.eu/growth/single-market/barriers-to-trade/tris_en.

The Guardian, 2017. Kenya brings in world's toughest plastic bag ban: four years jail or $40,000 fine. Available from: https://www.theguardian.com/environment/2017/aug/ 28/kenya-brings-in-worlds-toughest-plastic-bag-ban-four-years-jail-or-40000-fine [Accessed 28 August 2017].

The Independent, 2011. Myanmar's main city bans plastic bags: State media, 2011. Available from: http://www.independent.co.uk/environment/myanmars-main-city-bans-plastic-bags-state-media-2275309.html.

The Straits Times, 2017. Ban on non-biodegradable plastic bags takes effect in Malaysia Federal Territories, 2017. Available from: http://www.straitstimes.com/asia/se-asia/ ban-on-non-biodegradable-plastic-bags-takes-effect-in-malaysias-federal-territories.

The Summit Foundation, 2017. Eliminating plastic pollution on the Mesoamerican reef. Available from: http://www.summitfdn.org/mesoamerican-reef/eliminating-plastic-pollution-on-the-mesoamerican-reef/.

Tuoi Tre News, 2017. Vietnam considers fivefold gallop in plastic bag tax. Available from: https://tuoitrenews.vn/news/business/20170928/vietnam- considers-fivefold-gallop-in-plastic-bag-tax/41784. Html.

Twomey, D., 2013. Tas town says national ban should be in the bag. *Econews*. Available from: http://econews.com.au/28246/tas-town-says-national- ban-should-be-in-the-bag/.

Udasin, S., 2016. Plastic bag ban to take effect on January 1. *The Jerusalem Post*. Available from: http://www.globes.co.il/en/article-42-of-israelis-dont-buy-plastic-bags-1001184395.

United Nations Development Programme – United Nations Environment Programme (UNDP-UNEP) Poverty –Environment Initiative, 2015. Burkina Faso endorses law on sustainable development and bans non-biodegradable plastic bags. http://www.unpei.org/latest-news/burkina-faso-endorseslaw-on-sustainable-development-and-bans-nonbiodegradable-plastic-bags.

United Nations Development Programme (UNDP), 2015. Malawi introduces ban on thin plastic. Available from: http://www.mw.undp.org/content/malawi/en/home/presscenter/ articles/2015/09/08/malawi-introduces-ban-on-thin-plastic.html.

United Nations Development Programme (UNDP), 2017. Costa Rica paves the way to end single-use plastics. Available from: http://www.undp.org/content/undp/en/home/blog/ 2017/7/14/Costa-Rica-abre-el-camino-hacia-el-fin-de-los-pl-sticos-de-un-solo-uso.html.

United Nations Environment Programme (UNEP), 2017. Colombia's plastic bag tax: A concrete step towards fighting marine litter in the Caribbean, 26 July. http://web.unep.org/stories/story/colombia%E2%80%99s-plastic-bagtax-concrete-step-towards-fghting-marine-littercaribbean.

United Nations Environment Programme (UNEP), 2018. SINGLE-USE PLASTICS: A Roadmap for Sustainability (Rev. ed., pp. vi; 6).

Wakabi, M., 2015. Uganda in plastic bag ban dilemma. *The East African*. Available from: http://www.theeastafrican.co.ke/news/Uganda-in-plastic-bag-ban-dilemma/2558-2683982-4oll4l/index.html.

Watson, C., 2013. Plastic bag use still rife despite South Australia's shopping bag ban. Available from: http://www.adelaidenow. com.au/news/south-australia/plastic-bag-use-still-rife-despite-south-australias-shopping-bagban/newsastory/a02398d8295da04dcbe04b5343377186?sv=467d4722e2060125ce98c7a555a90d2f.

Wayang, L., 2017. Stop use of plastic bags. *National New*, 28 September. Available from: https://postcourier.com.pg/stop-use-plastic-bags/.

Wood Buffalo, Alberta, Canada, 2010. Single-use shopping bag bylaw. https://www.rmwb.ca/DoingBusiness/Bylaw-Enforcement/Single-Use-ShoppingBag-Bylaw.htm.

Xanthos, D., Walker, T. R., 2017. International policies to reduce plastic marine pollution from single-use plastics (plastic bags and microbeads): a review. *Marine Pollution Bulletin*, 118(1–2), 17–26.

Xuereb, M., 2009. Eco tax on plastic bags from March 1. *Times Malta*, 29 January. https://www.timesofmalta.com/articles/view/20090129/local/eco-tax-on-plastic-bags-from-march-1.242668.

Zen, I., Ahamad, R. B., Omar, W., 2013. No plastic bag campaign day in Malaysia and the policy implication. *Environment, Development and Sustainability*, 15. doi: 10.1007/s10668-013-9437-1.

Zitamar News, 2016. Mozambique plastic bag ban comes into force next Friday. Available from: https://zitamar.com/mozambique-plastic-bag-ban-comes-into-force-next-friday/.

Zohny, H., 2009. Red Sea Governorate bans plastic bags. *Egypt Independent*. Available from: http://www.egyptindependent.com/red-sea-governorate-bans-plastic-bags/.

Zoljargal, M., 2013. Less paper, more plastic. *The UB Post*. Available from: http://ubpost.mongolnews.mn/?p=3657.

3 Degradation Pathways of Various Plastics

Hyunjung Kim and Sadia Ilyas
Hanyang University

Gukhwa Hwang
Jeonbuk National University

CONTENTS

A change in the properties of a polymeric material like shape, tensile strength, color, and molecular weight under the influence of environmental factors such as light, heat, chemicals, or any other applied force is termed degradation. Chains containing aromatic functionality and epoxies are particularly susceptible to ultraviolet (UV) degradation, while hydrocarbon-based materials are susceptible to thermal degradation. The degradation of polymers into smaller molecules may proceed by random or specific scission. The processes by which polymers can suffer degradation are photolytic, thermal, mechanical, hydrolytic, chemical or biological, etc. depending on the type of polymer, morphology, molecular size, and the conditions to which it is subjected (Andrady et al., 2015; Gewert et al., 2015).

3.1 PHOTO-OXIDATIVE DEGRADATION

Most plastics are susceptible to photo-oxidative degradation that depends on the presence of UV absorbing species (chromophores).

Since saturated polyolefins cannot absorb much UV light directly but by chromophores of catalyst residues, pigments, flame retardants, processing aids, or double bonds containing organic molecules. These molecules release some of the absorbed UV energy by bonds breaking and releasing free radicals. Free radicals then begin a degradation cycle similar to autoxidation processes. Polyolefins are generally degraded by UV wavelengths of 290–300 and 310–340 nm (310–360 nm for poly-propylene). Without some sort of internal protection or stabilization, POs can lose properties relatively quickly under sunlight or UV exposure (Davis et al., 2010; Sherrington et al., 2016; Singh and Sharma, 2008). Degradation process steps can be summarized as follows:

1. UV energy is absorbed by chromophores, which leads to creating bond breaking and R$^{•}$ (free radicals) generation.
2. O_2 (oxygen) combines with R$^{•}$ to produce peroxy radicals (ROO$^{•}$) and hydro-peroxides (ROOH) along with some other species (H_2O, H_2, H_2O_2 (Eq. 3.1) (Here R=R$^{°}$).

$$O_2 + R^{°} \rightarrow ROO^{°} + RH \rightarrow ROOH + R^{°} \tag{3.1}$$

3. The hydroperoxides (ROOH) react themselves and produce new free radi-cals like alkoxy (RO$^{•}$) and hydroxyl ($^{•}$OH) radicals (Eq. 3.2).

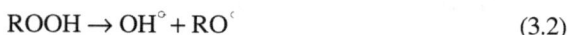

$$ROOH \rightarrow OH^{°} + RO^{°} \tag{3.2}$$

Aided by UV-assisted photo-oxidation, the degradation process results in cross-linking reactions or chain scission and is predominant in polyethylene and polypropylene (Figure 3.1).

Due to the presence of unsaturated (C=C) bonds, hydroperoxide, carbonyl, and hydroxyl groups in polymer chains, the poly(vinyl chloride) has poor light stability in the wavelength range of 253–310 nm. The chromophore-initiated photodegradation of poly(-vinyl chloride) is primarily determined by two factors: its ability to absorb UV light in the wavelength range under consideration and its participation by forming active particles (radicals) that further degrade the polymer chain. The terminal and internal alkene unsat-urated (C=C) bonds cannot be the primary initiators of poly(vinyl chloride) photodegra-dation under the action of sunlight with ($\lambda > 250$) nm because they absorb only UV light at ($\lambda < 200$) nm. The absorption by conjugated (C=C) bonds such as dienes and trienes shifts toward the longer wavelengths (Law, 2017; Yang et al., 2018; Wang et al., 2020).

The exposure of vinyl chloride polymers to light in the range of 250–350 nm develops the following signs of degradation (Yousif et al., 2012):

1. Discoloration from natural to dark brown or black.
2. Appearance of cracks on the surface.
3. Polymeric material can get soft and brittle.
4. Mechanical properties are altered like tensile strength, impact strength, elongation, and elasticity.
5. Transparency of material can change.
6. A deposit formation can occur on a material surface.

Under the influence of UV light (in the presence of air), rapid discoloration and grad-ual embrittlement start in polystyrene. Absorption of quantum of light (by polymeric molecule or impurity) leads to free radical formation.

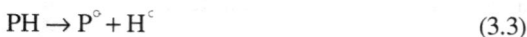

$$PH \rightarrow P^{°} + H^{°} \tag{3.3}$$

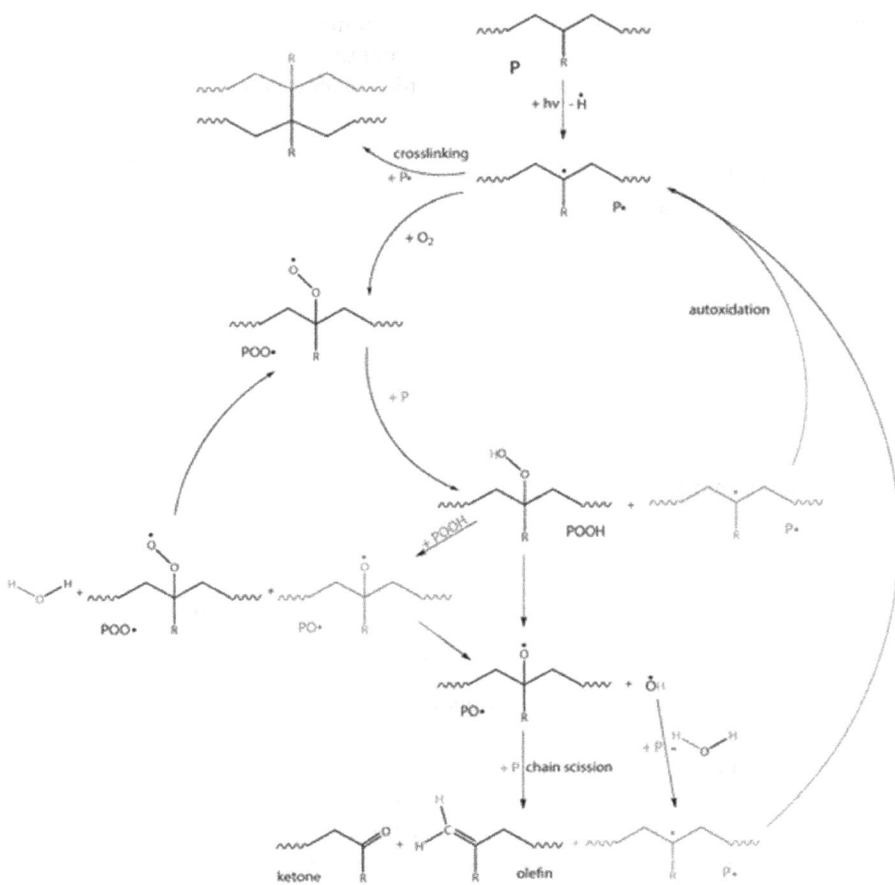

FIGURE 3.1 Various degradations in polyolefin. Here PE: polyethylene (R=H), PP: polypropylene (R=CH$_3$), PS: polystyrene (R=aromatic ring), and P: polymer backbone after photolytic cleavage of a C–H. (Retrieved from Gewert et al., 2015 under a Creative Commons License Copyright Elsevier.)

Pure polystyrene is considered stable above 290 nm but the phenyl group of polystyrene absorbs light when irradiated well below 280 nm. During this initiation, benzene rings excite to singlet and triplet states that are followed by reactions giving free radical formation by bond scission.

Complexes resulting from intra- or intermolecular charge transfer (CTCs) can form between electron-donating and electron-accepting groups, e.g. molecular oxygen and phenyl groups in polystyrene.

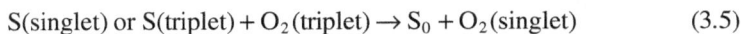

$$S + hv \rightarrow S(\text{singlet}) \rightarrow S(\text{triplet}) \tag{3.4}$$

$$S(\text{singlet}) \text{ or } S(\text{triplet}) + O_2(\text{triplet}) \rightarrow S_0 + O_2(\text{singlet}) \tag{3.5}$$

Singlet oxygen can be formed by energy transfer from the excited triplet state of phe-nyl groups, from CTCs and impurities (internal impurities; modified groups, external impurities; added groups) or external impurities (added compounds) to molecular oxygen.

Carbonyl groups can absorb near UV (290–400 nm) to form carbonyl biradicals that can abstract hydrogen from the same or neighboring macromolecules.

Hydroperoxide groups can decompose thermally or photochemically due to the low bond strength of the 0–0 bond.

The propagation of free radicals (presence of oxygen) can form peroxy radicals and hydro-peroxide groups.

$$R^{\circ} + O_2(\text{triplet}) \rightarrow ROO^{\circ} \tag{3.6}$$

The chain branching in photo-oxidation of polystyrene includes decomposition of hydro-peroxides into free radicals. The new radicals formed can take part in hydroxyl and carbonyl group formation on polymer chains.

The IR spectra show two complex absorption regions, at 1,800–1,600 cm^{-1} and 3,600–3,400 cm^{-1} of carbonyl and hydroxyl (or hydroperoxy) groups (Figure 3.2) (Rnby and Lucki, 1980).

The two 1R bands (in photo-oxidized polystyrene) have several absorption peaks indicative of different hydroxyl and carbonyl groups by various mechanistic routes simultaneously. Hydroxyl groups are formed along the main chain of polystyrene by the reaction of alkoxy polystyrene radicals with other polystyrene molecules.

A Cleavage of alkoxy radicals, an intermediate of photooxidation of 2-phenyl-butane, leads to the formation of acetophenone and propiophenone.

The termination of polystyrene chain radicals is due to a mutual combination that leads to the formation of cross-linked inactive products are formed. The new bonds can be either peroxides (at high oxygen pressure) or C–C bonds (at low oxygen pres-sure). The neighboring peroxide radicals on a chain can also have a probability to recombine and form peroxides or epoxides. Furthermore, it is noticed that scission and crosslinking may occur simultaneously (Figure 3.3).

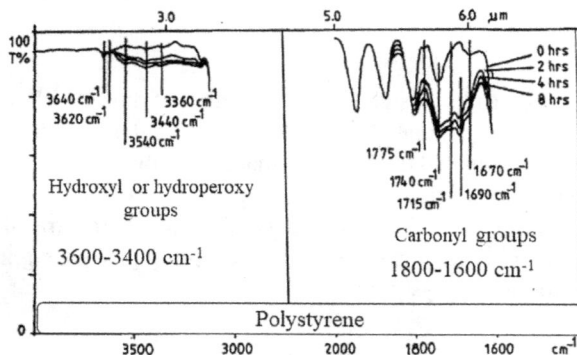

FIGURE 3.2 Infrared absorption spectra commercial polystyrene (irradiation 253.7 nm, presence of air). (Reproduced from Rnby and Lucki, 1980 after significant modifications.)

Stable cyclic peroxide formation by chain termination

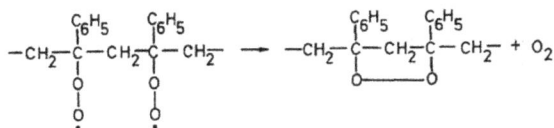

Stable cyclic epoxide formation by chain termination

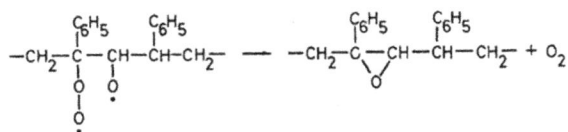

FIGURE 3.3 Chain termination in polystyrene. (Modified from Rnby and Lucki, 1980; Jellinek and Lipvac, 1970.)

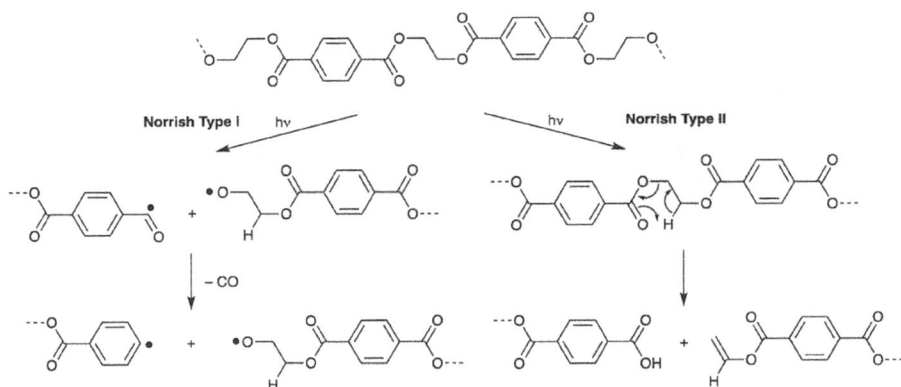

FIGURE 3.4 Formation of CO and COOH end-groups as results of Norrish type I and II reactions. (Adopted from Sang et al., 2020 after copyright permission.)

Polyethylene terephthalate (PET) contains ethylene glycolate and terephthalate subunits linked via ester bonds.

Under the influence of UV irradiation, the PET undergoes chain scission that occurs due to Norrish type I and type II which lead to the formation of free radicals (Figure 3.4).

Norrish type I leads to the formation of free radicals by ester bond cleavage, whereas Norrish type II are mainly intramolecular reactions. As a result of intramolecular photolysis, a hydrogen atom is abstracted and chains are terminated by forming carboxylic acid and vinyl end groups. The vinyl end group further leads to the formation of acetaldehyde upon hydrolysis.

Day and Wiles (1971) proposed the formation of carboxylic acid end groups in Norrish type II reactions via a cyclic intermediate rather than with oxygen. Fechine et al. (2004) proposed the formation of carboxylic acids, aldehydes, and hydroxyl terephthalate during photolysis by an alternative Norrish type I mechanism (Fechine et al., 2004; Sang et al., 2020).

3.2 THERMAL DEGRADATION OF PLASTICS

Thermal degradation of polymeric material is molecular deterioration under the influence of heat that can be further non-oxidative or oxidative.

Three fundamental reactions during non-oxidative thermal degradation include de-polymerization, random chains scission, and side-group elimination. In some cases, a recombination process that leads to cyclization or cross-linking can occur during degradation (Król-Morkisz and Pielichowska, 2019).

De-polymerization is considered a free radical mechanism where polymer chains are degraded to monomers, dimers, and/or oligomers at a high temperature. These reactions generally start at the chain ends where free radicals are formed (e.g. initiation). Then, the monomers are successively separated from the main chains during de-propagation. De-polymerization lasts till termination. Thermal de-polymerization is characteristic of some polymers obtained by chain polymerization like polystyrene, poly(methyl methacrylate), and polyoxymethylene.

During random chain scission, a typical degradation mechanism for polyolefins, the free radicals are randomly created along the main polymeric chains, which leads to the fragmentation of polymer into smaller molecules of different chain lengths (small enough to be volatile). Besides random scission, the end-chain scission process can be another possible pathway for the polymer thermal degradation in which monomers are effectively removed from the chain ends. The third degradation mechanism is side-group elimination in which detached side groups react with each other and convert into cyclic structures by the cyclization process (Król-Morkisz and Pielichowska, 2019).

Plastics can also undergo thermo-oxidative degradation at a high temperature. After absorption of sufficient heat the long polymer chains can break, which leads to the formation of free radicals (Pirsaheb et al., 2020; Peterson et al., 2001). These free radicals can react with oxygen and produce hydroperoxide. This self-propagation continued until the formation of inert products by a radical's collision and discontinuation of energy supply.

The temperature at which oxidative thermal degradation is initiated depends on the type of material and availability of oxygen. A glass transition temperature (T_g) is the temperature at which a polymer turns from a hard glassy state to a flexible soft state (Crawford and Quinn, 2016; Kotoyori, 1972; Rudin and Choi, 2013).

Regarding the maximum service temperature limit, the value depends on semicrystalline or amorphous nature of the material. A semicrystalline material can absorb a given amount of heat before it quickly changes from a solid to a low viscosity liquid. Thus, the most important temperature that determines the maximum service temperature is a melting point for these materials. However, amorphous plastics tend to get softer as they are heated indicating glass transition temperature as an important service temperature of these materials.

As a result of prolonged exposure to the radiative heat of the sun, softening can also be induced in many plastic materials, thereby resulting in morphological changes and affecting their resistance to mechanical degradation (Kamweru et al., 2011; Andrady, 2015).

3.3 CHEMICAL DEGRADATION OF PLASTICS

The presence of pollutants such as sulfur dioxide (SO_2), ozone (O_3), nitrogen dioxide (NO_2), and volatile organic compounds (VOCs), from atmosphere, can either degrade plastic materials directly or by catalyzing the formation of free radicals by photochemical reactions (Crawford and Quinn, 2016).

UV and lightening can produce ozone naturally but at low concentration. However, the presence of NO_x, SO_2, and VOCs in the air can increase ground O_3 level. (Placet et al., 2000). O_3 attacks the polymer's unsaturated C=C double bonds even at low concentrations and causes chain scission. O_3 also reacts with saturated polymers, but at a much slower rate (Cheremisinoff 2001).

SO_2 can be excited by UV irradiation, producing a reactive singlet or triplet state that reacts with unsaturated C=C double bonds either directly or by producing O_3 via photochemical reaction with O_2. NO_2 is very reactive due to the existence of odd electrons in the molecule, which can easily react with the unsaturated C=C double bonds, and the photochemical reaction of NO_2 with O_2 also produces ozone (McKeen, 2019).

In water environments, the most important chemical factors influencing plastic degradation are salinity of the water and pH value. High concentrations of H^+ (acidic) or OH^- (basic) in an aquatic environment may be able to catalyze the degradation of plastics that are susceptible to hydrolysis such as polyamides (Hocker et al., 2014; Wadsö and Karlsson, 2013). These two factors can also alter the surface of other types of plastics and microplastics and influence their behavior in a water environment and toward other constituents and pollutants in the water (Liu et al., 2019).

3.4 MECHANICAL DEGRADATION OF PLASTICS

Under the influence of external forces, due to collision and abrasion with sands and rocks, the degradation of plastics is termed mechanical degradation. The freezing and thawing of plastics in aquatic environments can also initiate mechanical degradation (Pal et al., 2018).

The effect of the external forces like shear forces, tension, and/or compression depends on the mechanical properties of the plastics (Crawford and Quinn, 2016; Allen et al., 1992; De Falco et al., 2020; Cesa et al., 2020)

Mechanical forces are mostly responsible for generating brake wear, tire wear, and road wear particles (Wagner et al., 2018). The interactions between a tire and a road surface and a brake pad and a brake disk yield frictional stresses those can fragment off the rubber surface directly. Tire treads are subjected to continuous stress during driving, and the rubber is pressed into a bulge, which creates prolonged stretching and causes material fatigue. Stress generated by abrasion with a road surface can reach the limiting strength of the material, resulting in micro-cutting or scratching of the tire and creating elongated rubber particles (Sommer et al., 2018).

Chain scissions of polymers during the photo, thermal, and chemical degradation will affect the plastics' mechanical properties, particularly their tensile elongation at break and tensile modulus (Andrady et al., 2015). Degradation in the environment was found to be able to decrease the elongation at break values of the plastics (O'Brine and Thompson, 2010), which lowers the requirement of external forces for the fragmentation of plastics and facilitates the mechanical degradation of plastics.

3.5 ATMOSPHERIC OXIDATION AND HYDROLYTIC DEGRADATION

Atmospheric oxygen can catalyze the breakdown of some plastic materials termed atmospheric degradation.

The most significant damage occurs from highly reactive ozone in the atmosphere, which can attack the double bonds of some plastics and elastomers by ozonolysis. The ozonolysis leads to the interaction of ozone molecule reacts with the double bond to produce an unstable, reactive ozonide (polyatomic anion). The rapid decomposition of the ozonide results in double bond cleavage and ultimately polymer chain breaking. The chain breaking by ozonolysis leads to a decrease in molecular weight and a reduction in material strength that can further cause brittlement and material cracking. For example, in elastomers, ozone can induce cracking on surfaces exposed to the atmosphere, and this ozone-cracking effect is often seen in old car tires. However, elastomers like Neoprene have good ozone resistance because the double bonds in the polymer chain are protected from attack by ozone due to the presence of chlorine in polymer backbone which decreases the electron density in the double bonds, thereby reducing the tendency to react with ozone (Davis et al., 2010).

Similarly, when some plastic materials are submerged in water, diffusion of the water into the amorphous regions of the plastic occurs. This diffusion of water into the polymer matrix can result in the addition of water molecules to the polymer by way of the cleavage of chemical bonds (hydrolysis). For example, the polyester PET is hydrolyzed at temperatures above the glass transition temperature (73°C–78°C) in which a scission reaction, catalyzed by oxonium ions or hydrogen ions produced by the carboxyl end groups, breaks the primary bonds of the polymer chain resulting in irreversible damage. Other plastics that suffer the effects of moisture are polyurethanes (Crawford and Quinn, 2016).

3.6 BIOTIC DEGRADATION OF PLASTICS

Generally, conventional plastics have an extremely low bioavailability but the degradation of biodegradable plastics relies upon biological processes that utilize the carbon present in the plastic as an energy source. However, in order for a plastic to be able to degrade, it must undergo an indirect procedure. Firstly, the oxygen, moisture, heat, ultraviolet light, or microbial enzymes break the carbon–carbon bonds of the long polymer chains resulting in the fragmentation of the plastic. Once the polymer has fragmented sufficiently, the shorter carbon polymer chains are able to pass through microbial cell walls. The carbon in the chains is then utilized as a food and energy source by the microbes, before being converted to biomass, water, carbon dioxide, or methane gases (Battin et al., 2016; Crawford and Quinn, 2016).

Santo et al. (2013) indicated the role of laccase in the biodegradation of polyethylene by the actinomycete *Rhodococcus ruber.* Hydroquinone peroxidase was found to be responsible for the biodegradation of PS by *Azotobacter beijerinckii* (Nakamiya et al., 1997). It has also been suggested that several enzymes excreted by fungi are capable of decreasing the length of polyethylene polymer chains (Sánchez, 2020).

Hydrolyzable polymers such as PET, PA, and polyurethane are usually more susceptible to biodegradation due to the presence of existing biodegradation pathways such as extracellular hydrolases involved in the degradation of cellulose and proteins (Chen et al., 2019). PETase, an enzyme capable of hydrolyzing PET initially identified in *Ideonella sakaiensis* was found to be ubiquitous in the environment (Danso et al., 2018). Enzymes such as cutinase, lipase, serine esterase, and nitro-benzyl-esterase have also been found to be capable of hydrolyzing PET, whereas protease, cutinase, amidase, and hydrolase are involved in the hydrolysis of PA (Guebitz and Cavaco-Paulo, 2008). Meanwhile, esterase and polyester hydrolase from bacteria and fungi might be responsible for polyurethane hydrolysis (Akutsu et al., 1998; Russell et al., 2011). In addition to hydrolysis, enzymatic oxidation may also contribute to the oxidative degradation of hydrolyzable polymers (Magnin et al., 2020).). Non-hydrolyzable polymers can be oxidized by O_2 with the catalysis of those enzymes, resulting in the formation of degradation products of low molecular weight.

Although biodegradable plastics may appear to be the perfect solution to plastic waste in the environment; in reality, this is not the case. The main difficulty with biodegradable plastics is that they are unpopular with manufacturers compared to the highest-produced commodity thermoplastics in terms of cost, versatility and mechanical strength, etc. Figure 3.5 represents various biotic degradation pathways after abiotic initiation.

ACKNOWLEDGMENTS

This work was supported by a grant from the National Research Foundation of Korea (NRF) grant funded by the Korea government (MSIT) (No. NRF-2020R1A2C1013851).

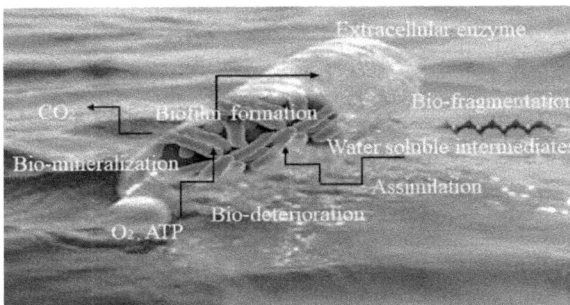

FIGURE 3.5 Biotic degradation for various plastics.

REFERENCES

Allen, G., Aggarwal, S. L., Russo, S. 1992. *Comprehensive Polymer Science: Supplement*, vol. 2, Pergamon Amsterdam Press, The Boulevard, Langford Lane, Kidlington.

Andrady, A.L., Bergmann, M., Gutow, L., Klages, M. 2015. *Marine Anthropogenic Litter*, Springer International Publishing, Cham. ISBN 978-3-319-16509-7.

Akutsu, Y., Nakajima-Kambe, T., Nomura, N., Nakahara, T. 1998. Purification and properties of a polyester polyurethane-degrading enzyme from Comamonas acidovorans TB-35. *Applied and Environmental Microbiology*, 64(1), 62–67.

Battin, T. J., Besemer, K., Bengtsson, M. M., Romani, A. M., Packmann, A. I. 2016. The ecology and biogeochemistry of stream biofilms. *Nature Reviews Microbiology*, 14(4), 251–263.

Cesa, F. S., Turra, A., Checon, H. H., Leonardi, B., Baruque-Ramos, J. 2020. Laundering and textile parameters influence fibers release in household washings. *Environmental Pollution*, 257, 113553.

Crawford, C. B., Quinn, B. 2016. *Microplastic Pollutants*, Elsevier. ISBN 978-0-12-809406-8.

Cheremisinoff, N. P. (Eds.) 2001. *Condensed Encyclopedia of Polymer Engineering Terms*, Butterworth-Heinemann, Boston, MA.

Chen, X., Xiong, X., Jiang, X., Shi, H., Wu, C. 2019. Sinking of floating plastic debris caused by biofilm development in a freshwater lake. *Chemosphere*, 222, 856–864.

Davis, M. E., Zuckerman, J. E., Choi, C. H. J., Seligson, D., Tolcher, A., Alabi, C., A., Yen, Y., Heidel, J. D., Ribas, A. 2010. Evidence of RNAi in humans from systemically administered siRNA via targeted nanoparticles. *Nature*, 464(7291), 1067–1070.

Danso, D., Schmeisser, C., Chow, J., Zimmermann, W., Wei, R., Leggewie, C., Streit, W. R. 2018. New insights into the function and global distribution of polyethylene terephthalate (PET)-degrading bacteria and enzymes in marine and terrestrial metagenomes. *Applied and Environmental Microbiology*, 84(8), e02773–17.

Day, M., Wiles, D. M. 1971. Photochemical decomposition mechanism of poly(ethylene terephthalate). *Journal Polymer Science, Part C: Polymer. Letters*, 9(9), 665–669. doi: 10.1002/pol.1971.110090906.

De Falco, F., Cocca, M., Avella, M., Thompson, R. C. 2020. Microfiber release to water, via laundering, and to air, via everyday use: A comparison between polyester clothing with differing textile parameters. *Environmental Science & Technology*, 54(6), 3288–3296.

Fechine, J. M., Rabello, M. S., Maior, R. S., Catalani, L. H. 2004. Surface characterization of photodegraded poly(ethylene terephthalate). The effect of ultraviolet absorbers. *Polymer*, 45(7), 2303–2308. doi: 10.1016/j.polymer.2004.02.003.

Gewert, B., Plassmann, M.M., MacLeod, M. 2015. Pathways for degradation of plastic polymers floating in the marine environment. *Environmental Science: Processes & Impacts*, 17, 1513–1521.

Guebitz, G. M., Cavaco-Paulo, A. 2008. Enzymes go big: Surface hydrolysis and functionalisation of synthetic polymers. *Trends in Biotechnology*, 26(1), 32–38.

Hocker, S., Rhudy, A.K., Ginsburg, G., Kranbuehl, D.E. 2014. Polyamide hydrolysis accelerated by small weak organic acids. *Polymer* 55 (20), 5057–5064.

Jellinek, H.H.G., Lipvac, S.N. 1970. Diffusion controlled oxidative degradation of isotactic polystyrene at elevated temperatures.*Macromolecules*, 3, 237.

Król-Morkisz, K., Pielichowska, K. 2019. *Polymer Composites with Functionalized Nanoparticles*. Elsevier, Amsterdam, The Netherlands.

Kamweru, P. K., Ndiritu, F. G., Kinyanjui, T. K., Muthui, Z. W., Ngumbu, R. G., Odhiambo, P. M. 2011. Study of temperature and UV wavelength range effects on degradation of photo-irradiated polyethylene films using DMA. *Journal of Macromolecular Science, Part B*, 50(7), 1338–1349.

Kotoyori, T. 1972. Activation energy for the oxidative thermal degradation of plastics. *Thermochimica Acta*, 5(1), 51–58.

Law, K. L. 2017. Plastics in the marine environment. *Annual Review of Marine Science*, 9, 205–229.

Liu, F. F., Liu, G. Z., Zhu, Z. L., Wang, S. C., Zhao, F. F. 2019. Interactions between microplastics and phthalate esters as affected by microplastics characteristics and solution chemistry. *Chemosphere*, 214, 688–694.

Magnin, A., Pollet, E., Phalip, V., Avérous, L. 2020. Evaluation of biological degradation of polyurethanes. *Biotechnology Advances*, 39, 107457.

McKeen, L. W. 2019. *The Effect of UV Light and Weather on Plastics and Elastomers*, William Andrew, Norwich, New York, US.

Nakamiya, K., Sakasita, G., Ooi, T., Kinoshita, S. 1997. Enzymatic degradation of polystyrene by hydroquinone peroxidase of Azotobacter beijerinckii HM121. *Journal of Fermentation and Bioengineering*, 84(5), 480–482.

O'Brine, T., Thompson, R. C. 2010. Degradation of plastic carrier bags in the marine environment. *Marine Pollution Bulletin*, 60(12), 2279–2283.

Pal, P., Pandey, J.P., Sen, G. 2018. In: Thakur, V.K. (Ed.), *Biopolymer Grafting*. Elsevier, Amsterdam, pp. 153–203.

Peterson, J. D., Vyazovkin, S., Wight, C. A. 2001. Kinetics of the thermal and thermo-oxidative degradation of polystyrene, polyethylene and poly (propylene). *Macromolecular Chemistry and Physics*, 202(6), 775–784.

Pirsaheb, M., Hossini, H., Makhdoumi, P. 2020. Review of microplastic occurrence and toxicological effects in marine environment: Experimental evidence of inflammation. *Process Safety and Environmental Protection*, 142, 1–14.

Placet, M., Mann, C. O., Gilbert, R. O., Niefer, M. J. 2000. Emissions of ozone precursors from stationary sources: A critical review. *Atmospheric Environment*, 34(12–14), 2183–2204.

Rnby, B., Lucki, J. 1980. New aspects of photodegradation and photo-oxidation of polystyrene. *Pure Applied Chemistry*, 52, 295–303.

Rudin, A., Choi, P. (Eds.) 2013. *The Elements of Polymer Science & Engineering*. Academic Press, Boston, MA.

Russell, J. R., Huang, J., Anand, P., Kucera, K., Sandoval, A. G., Dantzler, K. W., Strobel, S. A. 2011. Biodegradation of polyester polyurethane by endophytic fungi. *Applied and Environmental Microbiology*, 77(17), 6076–6084.

Sang, T., Christopher, J. W., Gavin, H., George, J.P. B. 2020. Polyethylene terephthalate degradation under natural and accelerated weathering conditions. *European Polymer Journal*, 136, 109873. ISSN 0014-3057, doi: 10.1016/j.eurpolymj.2020.109873.

Santo, M., Weitsman, R., Sivan, A. 2013. The role of the copper-binding enzyme–laccase–in the biodegradation of polyethylene by the actinomycete Rhodococcus ruber. *International Biodeterioration & Biodegradation*, 84, 204–210.

Sánchez, C. 2020. Fungal potential for the degradation of petroleum-based polymers: An overview of macro- and microplastics biodegradation. *Biotechnology Advances*, 40, 107501.

Sherrington, C., Darrah, C., Hann, S., Cole, G., Corbin, M. 2016. Study to support the development of measures to combat a range of marine litter sources. Report for European Commission DG Environment. https://www.eunomia.co.uk/reports-tools/study-to-support-the-development-of-measures-to-combat-a-range-of-marine-litter-sources/.

Singh, B., Sharma, N. 2008. Mechanistic implications of plastic degradation. *Polymer Degradation and Stability*, 93, 561–584.

Sommer, F., Dietze, V., Baum, A., Sauer, J., Gilge, S., Maschowski, C., Gieré, R. 2018. Tire abrasion as a major source of microplastics in the environment. *Aerosol and Air Quality Research*, 18(8), 2014–2028.

Torikai, A., Takeuchi, A., Nagaya, S., Fueki, K. 1986. Photodegradation of polyethylene: Effect of crosslinking on the oxygenated products and mechanical properties. *Polymer Photochemistry*, 7(3), 199–211.

Wadsö, L., Karlsson, O. J. 2013. Alkaline hydrolysis of polymers with ester groups studied by isothermal calorimetry. *Polymer Degradation and Stability*, 98(1), 73–78.

Wagner, S., Hüffer, T., Klöckner, P., Wehrhahn, M., Hofmann, T., Reemtsma, T. 2018. Tire wear particles in the aquatic environment-a review on generation, analysis, occurrence, fate and effects. *Water Research*, 139, 83–100.

Wang, C., Xian, Z., Jin, X., Liang, S., Chen, Z., Pan, B., Gu, C. 2020. Photo-aging of polyvinyl chloride microplastic in the presence of natural organic acids. *Water Research*, 183, 116082.

Yang, H., Ma, M., Thompson, J. R., Flower, R. J. 2018. Waste management, informal recycling, environmental pollution and public health. *Journal of Epidemiology and Community Health*, 72(3), 237–243.

Yousif, E., Ahmed, A., Mahmoud, M. 2012. *New Organic Photo-Stabilizers for Rigid PVC Against Photodegradation*, LAP, LAMBERT, Germany.

4 Interaction of Inorganic and Organic Pollutants with Microplastics

Hyunjung Kim and Sadia Ilyas
Hanyang University

Humma Akram Cheema
University of Agriculture Faisalabad (UAF)

CONTENTS

DOI: 10.1201/9781003200628-4

MPs can efficiently serve as a vector of several contaminants (heavy metals, organic substances) and transport and migrate in different environments due to the large specific surface area, surface functional groups, and aging (Zhou et al., 2019; Holmes et al., 2014). Most common examples of potential pollutants are manganese (Mn), iron (Fe), zinc (Zn), aluminum (Al), copper (Cu), lead (Pb), silver (Ag) as well as hydrophobic organic pollutants, e.g., polychlorinated biphenyls (PCBs), poly-aromatic hydrocarbons (PAHs), and organochlorine pesticides. The primary mechanisms by which MPs adsorb pollutants include electrostatic interactions, hydrophobic interactions, partition, and other non-covalent interactions (Wang et al., 2015, Zhang et al., 2018a, b). During the adsorption process, multiple adsorption mechanisms take place in parallel and several factors affect these interactions.

4.1 INTERACTION OF INORGANIC POLLUTANTS WITH MICROPLASTICS

Heavy metals are a class of elements, which are characterized by their higher atomic weight and numbers (>20) as well as their higher densities (>5 g cm^{-3}), e.g., Pb, Cd, Zn, AS, Cu, Hg, Ni, Fe, Pt, Cr, and Pd. In nature, both natural and anthropogenic activities are the major source of heavy metals. Heavy metals are considered a potential pollutant in aqueous systems due to their non-degradable nature. MPs have a large surface area and have the ability to interact with heavy metals and concentrate them at a high level (Hansen et al., 2013; Holmes et al., 2012). Heavy metals can interact with MPs directly (electrostatic interaction, complex formation, bioaccumulation, precipitation) or indirectly (via extracellular polymeric substances. Figure 4.1 depicts various interaction mechanisms between inorganic pollutants and microplastics.

4.1.1 Mechanism of Interaction of Inorganic Pollutants with Microplastics

4.1.1.1 Electrostatic Interactions

Generally, electrostatic interaction occurs at a specific pH (Wang et al., 2020). If the pH of the adsorption environment exceeds the point of zero charges of the MPs, their surface will be negatively charged and electrostatically attract positively charged pollutants. However, when the pH of the adsorption environment exceeds the acid dissociation constant of the pollutant (mostly in the case of organic pollutants), they will be deprotonated and exist in an anionic form, causing electrostatic repulsion and inhibiting adsorption by MPs (Wu et al., 2019). Therefore, electrostatic interaction is closely related to the electrification of MPs, the form of the pollutants, and the quantity of charge involved. The polarity of the microplastic surface may originate from its physical and chemical properties. For example, polyvinylchloride (PVC) and chlorinated polyethylene contain chlorine and hexabromocyclododecane (serve as brominated flame retardant) has charged additives and impurities. Moreover, photo-oxidative weathering generates new adsorption bands (C=C, C–O, –OH) which also enhanced the polarity of the polymer and induce charged surface (Mato et al., 2001).

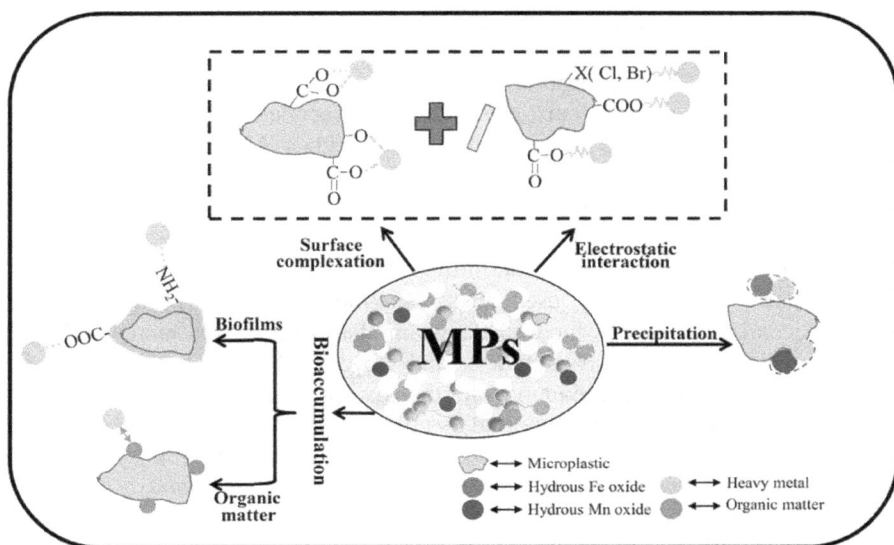

FIGURE 4.1 Schematics of various interaction mechanisms between inorganic pollutants and microplastics. (Adopted after permission from Cao et al., 2021.)

4.1.1.2 Sorption, Bioaccumulation, Complexation, and Co-precipitation

Inorganic pollutants can also interact with MPs via Sorption, bioaccumulation, co-precipitation, and complexation. Sorption is a physical and chemical process by which one substance becomes attached by adsorption, absorption, or ion exchange. Adsorption is the adhesion of atoms, ions, or molecules from a gas, liquid, or dissolved solid to a surface. This process creates a film of the *adsorbate* on the surface of the *adsorbent*. This process differs from absorption, where fluid (*absorbate*) is dissolved or permeates a liquid or solid (*absorbent*). Adsorption is a *surface phenomenon*, while absorption involves the whole volume of the material, although adsorption does often precede absorption. The term *sorption* encompasses both processes, while *desorption* is its reverse. While Ion exchange is a reversible reaction, charged ions present in the solution are replaced by similarly charged ions present within the insoluble exchange material (Crini, 2010).

MPs form new complexes via sorption or bioaccumulation through biofilms and natural organic matter. The hydrophobicity of MPs is decreased due to the growth of biofilms, and the abundance of carboxyl and ketonic groups onto the surface of MPs is increased, which enhances the adsorption capacity of heavy metals (Tu et al., 2020). Existing literature revealed that the biofilms formation might positively affect the adsorption of heavy metals, and the concentration of heavy metals on MPs will increase as the biofilm matures. Yet, it must be noted that during the whole biofilm growth, the long-term dynamic changes of heavy metals loading on microplastics remain largely unknown, and more work need to be done in this direction (Qi et al., 2021).

Heavy metallic cations and their complexes also co-precipitate with iron and manganese hydrous oxides through adsorption onto beached pellets (Ashton et al., 2010).

Dong et al. (2019) use a computational study to confirm the binding of As^{3+} with the carboxylic group present on the surface of PS-MPs polystyrene-microplastics through H-bonding. This adsorption mechanism comprises both electrostatic as well as non-covalent interactions. It also revealed that extracellular polymeric substances perform a major role in the biosorption of heavy metals. More extracellular polymeric substances can be produced from microbial cells in biofilms in the presence of heavy metals to protect biofilms from harsh environments (Dong et al., 2019).

The rate of heavy metal (inorganic pollutants) adsorption by MPs can be described by adsorption kinetics, and fitting of those can further indicate adsorption mechanisms. Frequently used adsorption models contain pseudo-first and pseudo-second-order kinetic models, Bangham channel diffusion models, Elovich kinetic models, Weber–Morris models, and Boyd models. A pseudo-first-order kinetic model was usually fitted model for the adsorption process (Zon et al., 2018); however, some researchers also reported that the pseudo-second-order kinetic model could yield a better fit (Nethaji et al., 2013; Guo et al., 2020; Tang et al., 2020). This model accepts that the adsorption of the heavy metals by MPs is primarily controlled by the chemisorption mechanism, which involves the either sharing or transfer of electron pairs. Still, it is not under control by the material transport step. The multistage nature of the adsorption process was described by using Weber–Morris model (Nethaji et al., 2013; Zon et al., 2018; Öz et al., 2019). According to this model, adsorption of heavy metals by MPs may proceed in multiple steps. Studies by Guo et al. (2020) further observed that heavy metals' adsorption onto MPs can be divided into three steps. The first step comprises the rapid interaction between heavy metal ions with active functional groups on the surface of MPs, primarily due to van der Waals and covalent forces. After saturation of these binding processes, the heavy metals start to diffuse on the surface pores of MPs. In the third and last step, the adsorption rate decreases significantly and ultimately establishes an equilibrium between the adsorption and desorption process.

The distribution of pollutants between the solid and liquid phases at equilibrium can be represented by an adsorption isotherm. Generally, Langmuir isotherm model (Langmuir, 1918), as well as Freundlich isotherm model, are used. The Langmuir isotherm model assumes no interaction force between the adsorbed molecules, and only monolayer adsorption can be formed on the MP surface. In contrast, the Freundlich isotherm model is an empirical equation with no assumptions. According to the Freundlich isotherm model, multilayer adsorption occurs on a heterogeneous surface during the adsorption process, and pollutants molecules formerly occupied high energy adsorption sites then diffuse to low energy adsorption sites (Abdurahman et al., 2020; Wang et al., 2020). Several studies demonstrate that both Langmuir and Freundlich models can effectively describe adsorption isotherms (Zon et al., 2018; Dong et al., 2020) (Table 4.1).

4.1.2 FACTORS EFFECTING THE INTERACTION OF INORGANIC POLLUTANTS WITH MICROPLASTICS

The interaction process is mostly influenced by the type and characteristics of plastics, the chemical properties of inorganic pollutants, and several other environmental

TABLE 4.1

Reported Adsorption Isotherm and Kinetics Model of Heavy Metals on Microplastics

MP Type	MP Size	Media Type	Types of Heavy Metals	Best-Fit Isotherm Model	Best-Fit Kinetics Model	References
Virgin and beached PE pellets	–	Filtered seawater	Cr, Co, Ni, Cu, Zn, Cd and Pb	Langmuir, Freundlich	The PFO model	Holmes et al. (2012)
Virgin and beached (aged) pellets (PE)	≤1 mm	River water and sea water	Cd, Co, Cr, Cu, Ni, Pb	Langmuir, Freundlich	–	Holmes et al. (2014)
Virgin HDPE MPs	1–2 mm, 0.6–1 mm, and 100–154 mm	In aqueous medium	Cd	Langmuir	The PSO model	Wang et al. (2019 a,b)
HDPE pieces	$0.92 \pm 1.09 \text{ mm}^2$ ($n=314$)	An arable and woodland soil	Zn	Freundlich	–	Hodson et al. (2017)
Virgin HDPE-MPs	48–58 μm, 100–154 μm, 0.6–1.0 mm, and 1.0–2.0 mm	A typical farmland	Cd	Langmuir	The PSO model	Zhang et al. (2020)
CPE, PVC, HPE and LPE	280 μm (60 meshes)	In aqueous medium	Cu^{2+}, Pb^{2+}, Cd^{2+}	Freundlich	–	Zou et al. (2020)
Collected nylon MPs	2 mm × 0.24 mm	In aqueous medium	Cu(II), Ni(II), Zn(II)	Langmuir, Freundlich	The Elovich model, The PSO model	Tang et al. (2020)
Natural-aged MPs, mostly PE	<5 mm	In aqueous medium	Pb(II)	Langmuir	–	Fu et al. (2020)
Virgin PA, PS, and PP	100–150 μm	In aqueous medium	Sr^{2+}	Nonlinear Temkin model	–	Guo et al. (2020)
Virgin and aged nylon MPs, PE	2 mm	In aqueous medium	Pb(II)	Langmuir	The PSO model	Tang et al. (2018)
Virgin PSMPs	0.1–1, 1–10, and 10–100 μm	In aqueous medium	As(III)	Langmuir, Freundlich	The PSO model	Dong et al. (2019)

(Continued)

TABLE 4.1 (Continued)

Reported Adsorption Isotherm and Kinetics Model of Heavy Metals on Microplastics

MP Type	MP Size	Media Type	Types of Heavy Metals	Best-Fit Isotherm Model	Best-Fit Kinetics Model	References
PVC, PE, PS	75–106 μm (150–200 mesh)	In aqueous medium	Pb(II)	PVC: Langmuir, Freundlich PS and PE: BET model	The PSO model	Lin et al. (2021)
Virgin PS beads and naturally aged PVC fragments	PS: 0.7–0.9 mm diameter; PVC: 1.6×0.8 mm in size	Seawater	Cu, Zn	–	The first-order approach to equilibrium model	Brennecke et al. (2016)
Virgin PS	0.6 mm	In aqueous solution	As, Pb	Langmuir	The PSO model	Zhou et al. (2019)
Virgin PE, PP, PVC and PS	75 μm (200-mesh)	In aqueous medium	Cd²⁺	Henry, Freundlich	The PSO model	Guo et al. (2019)
virgin PET debris and aged PET debris	About 1 mm × 1 mm	In aqueous solution	Cu2+, Zn2+	Langmuir	–	Wang et al. (2020)
New and weathered LDPE and PET MPs	0.5 cm × 0.5 cm	In the synthetic stormwater	Pb²⁺, Zn²⁺	–	For new LDPE MPS: the PFO model;For the weathered LDPE MPS: the PSO model	Aghilinasrollahabadi et al. (2021)
Virgin and aging PS	The average size is 150 μm	In aqueous solution	Cd²⁺	Freundlich	The PSO model	Chen et al. (2019)
Virgin PA, PVC, PS, ABS, PET	43–74 μm	In aqueous solution	Cd(II)	Freundlich	The PSO model	Zhou et al. (2019)

CPE; chlorinated polyethylene, PET; polyethylene terephthalate, ABS; acrylonitrile butadiene styrene, HDPE; high-density polyethylene, LDPE; low-density polyethylene, HPE; high crystallinity polyethylene, LPE; low crystallinity polyethylene

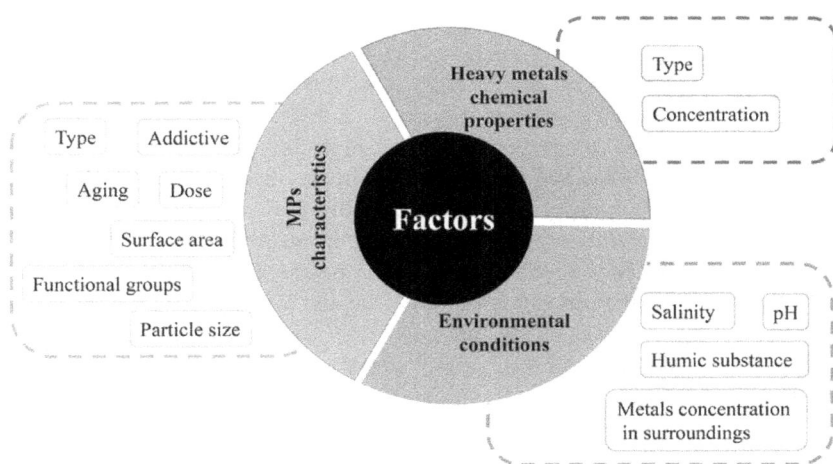

FIGURE 4.2 Factors affecting the adsorption behavior of heavy metals by various microplastics. (Adopted after permission from Cao et al., 2021.)

factors like natural organic matter, pH, salinity, and concentration of pollutants (Figure 4.2).

4.1.2.1 Effect of the Physicochemical Characteristics of Microplastics

Polypropylene (PP), polyethylene (PE), polystyrene (PS), and PVC are commonly studied polymers/MPs for the adsorption of heavy metals. The influence of polymer type mainly depends on surface area and functional group of MPs. For instance, Guo et al. (2019) studied the adsorption of Cd^{2+} ions by using four different MPs: PE, PP, PVC, and PS. Adsorption behavior (PVC > PS > PP > PE) was observed to be consistent with their specific surface area ($0.836 > 0.508 > 0.348 > 0.173\,m^3\,g^{-1}$) indicating a strong correlation between specific surface area and adsorption capacity of MPs. However, Lin et al. (2021) exhibited a different adsorption order (PS ($128.5\ \mu g\ g^{-1}$) < PE ($416.7\ \mu g\ g^{-1}$) < PVC ($483.1\ \mu g\ g^{-1}$)) that indicates the influential role of various other factors. Brennecke et al. (2016) studied the adsorption of copper ions in PVC and PS. They stated that PVC has a higher adsorption capacity than PS because it is more polar due to chlorine atoms and has a high surface area. Gao et al. (2019) also supported the above result. They declared that PVC has significantly higher adsorption capacity for various heavy metals such as lead, cadmium, and copper in comparison with other MPs like polyamide (PA), PE, and polyoxymethylene (Tang et al., 2020) suggested that oxygen-containing functional groups on the surface of MPs such as nylon also have effective adsorption capacity for several heavy metals ions such as zinc, copper, and nickel. Furthermore, the adsorption behavior of MPs was also tuned by the addition of plastic additives. Very recent research revealed that hexabromocyclododecane considerably improved the adsorption of lead (Pb^{2+}) on the composite of hexabromocyclododecane-polystyrene MPs, and from 0 to $0.760\ \mu mol\ g^{-1}$ lead was adsorbed on polystyrene (Lin et al., 2020).

Moreover, MPs dose and particle size also perform a major role in the adsorption process. Recent studies showed a decrease in adsorption rate with PP concentration higher than 0.1 mg L^{-1} (Gao et al., 2019; Fu et al., 2019) due to fewer adsorption sites for MP particles in metallic solution (Gao et al., 2019). A decrease in the adsorption capacity of MPs for several heavy metals like lead, cadmium, and copper with a large particle size was also observed by several researchers due to less available surface area (Gao et al., 2019; Wang et al., 2019 a, b; Zhang et al., 2019).

The effect of aging on adsorption is also an important factor that cannot be ignored. The adsorption capacity of aged MPs for cadmium, lead, zinc, and copper was found to be 1–5 times more than in virgin MPs by Guo and Wang (2019). Brennecke et al. (2016) also observed that aged PVC adsorbs more zinc and copper before establishing equilibrium. Holmes et al. (2012) also revealed that the magnitude of partition coefficient of chromium was one time higher in aged PE in comparison with virgin PE. This effect of aging can be due to weathering, crushing, abrasion, and surface oxidation by UV radiation. After the UV radiation, oxygen-containing functional groups such as ether (C–O–C) and a hydroxyl group (OH) are produced, bringing the strong complexing ability of metallic ions with MPs. Moreover, the production of more binding sites or modification of surface properties induced during the aging process by precipitation of organic matter and inorganic minerals on the surface of MPs (Brennecke et al., 2016; Wang et al., 2020).

4.1.2.2 Effect of Chemical Properties of Inorganic Pollutants

The types of heavy metals define the atomic number and surface valence state that ultimately lead to affect adsorption capacity and surface potential. Additionally, synergistic or competitive effects among metals also affect adsorption rate and capacity. The adsorption capacity of various MPs can be promoted or reduced in several coexisting metal solutions. In Pb, Cu, and Cd-coexisting solution systems (0.05 mg L^{-1}), the adsorption capacity of mixed solutions was lower than that in a single metallic solution implying a competitive adsorption phenomenon. The coexisting of heavy metals also has a promoting effect except for competitive behavior. A study showed that the affinity of PA to Pb was stronger in the solution with Cu and Pb coexisting than in a single solution (Gao et al., 2019). Fu et al. (2019) also observed similar behavior and declared that the coexistence of zinc ions boosted the adsorption of copper ions on polystyrene. Fu et al. (2020) also found the coexistence of Zn^{2+} promoted the adsorption of Cu^{2+} on PS. Besides, the adsorption capacity of MPs is enhanced with increasing initial Pb^{2+} concentration in the range of 2–15 mg L^{-1} (Fu et al., 2020), suggesting that the initial concentration of metals can also affect the adsorption capacity. Several researches show the same conclusions, such as plastic production pellets adsorbed Ag, Cd, Co, Cr, Cu, Hg, Ni, Pb, and Zn in the range of 0.0004–2.78 µg g^{-1} at 20 µg L^{-1} (Holmes et al., 2012).

4.1.2.3 Effect of Environmental Factors

Different environmental conditions were also correlated with the adsorption capacity of heavy metals. Holmes et al. (2014) observed that Cd, Co, Ni, and Pb adsorption increased with increasing pH in river water. The reason is that the functional groups on the MPs surface are de-protonated, which increases the electronegativity and adsorption sites on the MPs surface. Another explanation is that the zeta potential of MPs is negative when pH is greater than point of zero charges; the higher the pH value, the more negative charge on the surface will be. It may generate more electrostatic attraction to the metal cation. However, as pH increases, the heavy metal ions may not be conducive to adsorption due to the passivation of precipitation. For example, Pb is mainly present as Pb^{2+} at low pH, and $Pb(OH)_2$, $Pb(OH)^{3-}$ at high pH (Lin et al., 2021).

As for the temperature, the general opinion is that high temperatures will benefit the adsorption of heavy metals on MPs (Öz et al., 2019). The possible explanation for this observation is that the adsorption process is an endothermic reaction; thus, the spontaneity of the adsorption process may increase with the increase in temperature.

Salinity also significantly influences adsorption behavior between heavy metals and MPs in different ways. Firstly, the ionic strength of sodium ions (Na^+) prompted by sodium chloride (NaCl) may compete with cadmium ions (Cd^{2+}) for the adsorption sites. Similarly, chlorine ions (Cl^-) can detain cadmium ions by forming complexes like $CdCl^+$, $CdCl_2^0$, $CdCl^{3-}$, and $CdOHCl^0$ when they co-exist (Wang et al., 2019a). Electrostatic shielding to metal is also caused by salinity and also affects electrostatic adsorption behavior. Lin et al. (2021) studied the effect of ionic strength on the adsorption of Pb^{2+} in 0.01 M and 0.1 M NaCl. It was found that the adsorption of Pb^{2+} was inhibited at high ionic strength on the surface of the MPs. An increase in salinity causes the accumulation of MPs and adsorption sites to decrease. Conversely, the electrostatic screening arose between the positive Pb^{2+} and negative surface by salt (Guo et al., 2019; Lin et al., 2021).

Humic substances can prevent metal ions (e.g., Cu, Cd, Zn, Pb, Cr, and As) from adsorption on various adsorbents (Tang et al., 2020). Nevertheless, in the MP–metal–humic substance system, the effect of fulvic acid and humic acid mainly depends on the type of heavy metals. The high concentration of fulvic acid inhibited copper and zinc ion adsorption capacities on nylon since complexes (/FA2M, bidentate) between metal ions and fulvic acid are formed, which hinders the metals from adsorbing onto the MPs surface through the surface complexation. While the formation of /FANi+ may facilitate the binding of nickel ions on nylons (Tang et al., 2020). On the other hand, the negative charge of humic acid adsorbed onto MPs will facilitate cadmium ions sorption onto it by electrostatic synergy (Guo et al., 2019). Lastly, the concentration of metals in the environment may also affect the adsorption capacity of MPs. Mostly, the adsorption of heavy metals on MPs was found to be consistent with their concentration in the surrounding environment.

4.2 INTERACTION OF ORGANIC POLLUTANTS WITH MICROPLASTICS

Organic pollutants contain carbon covalently linked to atoms of other elements, most commonly hydrogen, oxygen, or nitrogen. Some organic pollutants are persistent as persistent organic pollutants (POPs), such as organochlorine pesticides and PCBs. POPs show resistance toward degradation via biochemical and photolytic processes and can bio-accumulate with probable adversarial effects on the ecosystem. POPs can be effortlessly migrated from their sources and re-concentrated in the

Perfluorooctanoic acid (PFOA)

Polychlorinated biphenyls (PCBs)

Perfluorooctane sulfonic acid (PFOS)

Ovalene; Polycyclic aromatic hydrocarbons

Polybrominated diphenyl ether

Orthophthalates (general structure)

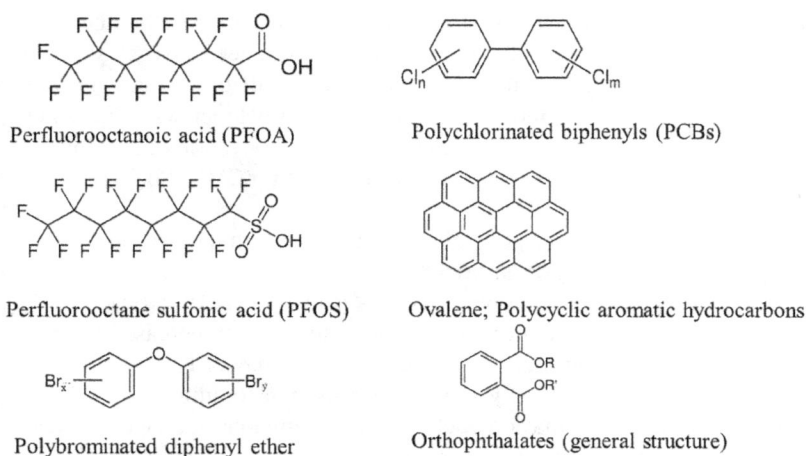

FIGURE 4.3 Representative structures of various organic pollutants.

new environment at possibly toxic levels. Both natural activities such as volcanoes and different biosynthetic pathways and artificial means such as the manufacturing of chemicals. They are lipophilic (tend to dissolve in fats and lipids) and have low solubility. Because of their low solubility, POPs also tend to associate with the suspended particles in the water column and accumulate on MPs. Consequently, in tandem with microplastics, the negative effects could be doubled (Verla et al., 2019). Commonly present organic pollutants that can interact with microplastics include Perfluoroalkylates, Phthalates, Polycyclic aromatic hydrocarbons, Polycyclic aromatic hydrocarbons, Polybrominated diphenyl ethers, and Polychlorinated biphenyls (Figure 4.3).

Perfluoroalkylated substances are a class of highly fluorinated chemicals commonly used in industries because of their beneficial surfactant properties and their capability to repel materials such as dirt, oil, and water. Therefore, they have extensive applications such as textiles, packaging materials, foams to extinguish fires, and paints. Perfluoroalkylated substances remain to persist in the aquatic system.

Phthalates are a class of chemicals that may add to plastics during manufacturing to make them softer and flexible (plasticize). Phthalates are used in PVC, which is used to make products such as plastic packaging, garden hoses, and medical tubing (https://www.cdc.gov/biomonitoring/Phthalates_FactSheet.html). Phthalates are also combined with MPs during manufacturing but do not chemically bond to plastic; hence they are easily leached out latterly (Titmus and Hyrenbach, 2011).

PAHs are a class of chemicals that can exist in more than 100 different forms and are considered as most ubiquitous pollutants in the environment.

PAHs are readily produced due to the incomplete combustion of wood, tobacco, and other fuel sources composed of carbon compounds (Pham et al., 2013). Consequently, it is recognized that some PAHs in the environment can originate from natural sources, such as forest fires and volcano eruptions (Endo et al., 2011).

However, their presence in the environment is mainly due to anthropogenic activities, such as coal-burning power plants and shipping activities (Webster et al., 2009). Furthermore, some PAHs are used in industry to produce plastics, pesticides, and dyes.

Ocean-based industrial oil-extraction platforms regularly emit PAHs into the atmosphere as part of their manufacturing process. PAH by-products from combustion can be washed into marine habitats via rainfall and watercourses or water settle from the atmosphere onto ocean surface waters. Moreover, it has been determined that the burning of plastic refuse emits PAHs, with polystyrene producing the highest quantities. Additionally, the manufacture of polystyrene can have PAHs as an undesired consequence of incomplete polymerization during processing in which the toxic PAH precursors, benzene and styrene, can become incorporated into the polymer matrix (Ryan et al., 2009).

Due to their toxicity and persistence, several families of brominated flame retardants have been listed as POPs in the Stockholm Convention, a multilateral treaty overseen by the United Nations Environment Programme. Polybrominated diphenyl ethers, mainly deca-BDE and octa-BDE, were added to plastic materials during manufacturing and acted as flame retardants (https://www.ices.dk/sites/pub/Publication%20Reports/Techniques%20in%20Marine%20Environmental%20Sciences%20(TIMES)/TIMES46.pdf).

PCBs are a class of synthetic chemical compounds consisting of biphenyl rings, and chlorine atoms are substituted at different positions of the benzene ring. Anthropogenic activities are one of the prime routes for the dispersion of PCBs in the atmosphere due to the vast application of these compounds in industry.

Due to their highly resilient nature and how PCBs distribute and behave in ecosystems, it has been estimated that PCBs will remain the most widespread contaminant of aquatic environments and organisms until at least 2050. The extent of chlorination directly affects the half-life of the different congeners in the atmosphere by dropping their liability to photo-degradation. Therefore, PCBs with a greater extent of chlorination have a tendency to persist more in the atmosphere, and the half-life of the several congeners has been predictable in the range of 10–548 days (Breivik et al., 2007).

4.2.1   Mechanism of Interaction of Organic Pollutants with Microplastics

4.2.1.1   Electrostatic Interactions

Organic pollutants can interact with MPs by electrostatic interactions. Electrostatic interactions comprise the attractive or repulsive interactions between charged molecules. Therefore, electrostatic interaction is closely related to the electrification of MPs, the form of the organic pollutants, and the quantity of charge involved (as discussed above).

Antibiotics are ionizable compounds, and their ionization constants vary depending on their functional groups. Antibiotics can exist as cations, zwitterions, or anions, depending on the pH (Wang et al., 2015). Guo et al. (2018) demonstrated that if the pH is below 7.1, TYL is positively charged while PS and PVC are negatively charged;

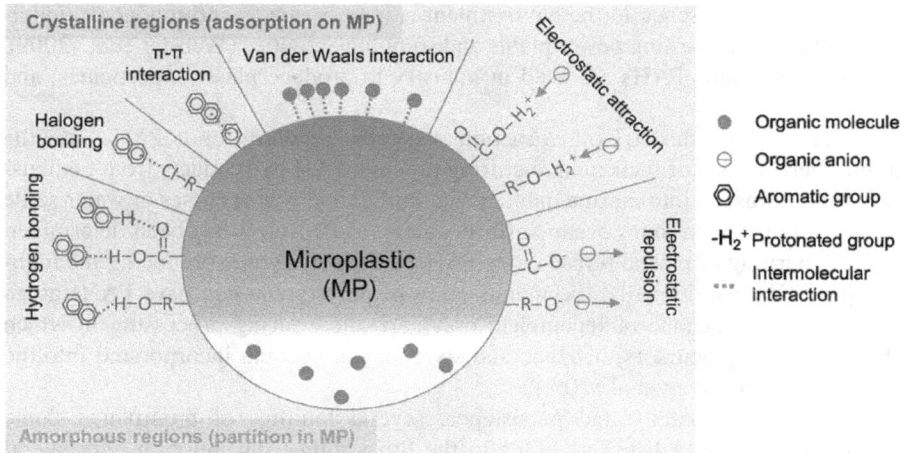

FIGURE 4.4 Mechanisms of the interactions between MPs and organic pollutants. (Adopted after permission from Mei et al., 2020.)

thus, the adsorption of TYL by these two MPs occurs electrostatically. In freshwater systems, SDZ, AMX, TC, and TMP exist in zwitterionic and anionic forms, while some CIP exists in cationic forms; hence, the CIP adsorption capacity of MPs is higher than those of the other four antibiotics. Additionally, the sulfamethoxazole adsorption capacity of PA in acidic media ($pH \leq 6.7$) is higher than that in alkaline media ($8 \leq pH \leq 9$). As the electronegativity of the MPs increases and the form of sulfamethoxazole changes to anionic in alkaline media, aggravating electrostatic repulsion between them and reducing the adsorption capacity (Guo et al., 2018; Guo and Wang, 2019). Figure 4.4 indicated the mechanisms of the interactions between MPs and organic pollutants

4.2.1.2 Hydrophobic Interactions and Partition Effect

Resin is the main component of plastics and has a non-wetting surface with strong hydrophobicity. Most organic pollutants are characterized by high-fat solubility and low water solubility; thus, organic pollutants are easily adsorbed onto the surface of MPs (Li et al., 2019). The octanol/water partition coefficient (K_{ow}/Log K_{ow}) represents the hydrophobicity of a substance (Zhang et al., 2012). Organic compounds with high Log K_{ow} values are likely to be absorbed by MPs more easily. Fang et al. (2019) observed that the amounts of three triazole fungicides, mold butyronitrile, and hexaconazole adsorbed by PS were consistent with their respective Log K_{ow} values. While studying the behavior of bisphenol adsorption on MPs, Wu et al. (2019) also observed a strong linear relationship between the equilibrium adsorption efficiency and the Log K_{ow} value, indicating that hydrophobic interaction is the primary mechanism by which MPs adsorb bisphenol. Furthermore, the adsorption of 9-nitroanthrene (9-NAnt) by PP, PE, and PS, perfluorooctane sulfonate (PFOS), and perfluorooctane sulfonamide by PE, and adsorption of 4-chlorophenol,

2,4-dichlorophenol, and 2,4,6-trichlorophenol by polyethylene terephthalate (PET) are also affected by hydrophobic interactions (Wang et al., 2015; Zhang et al., 2020; Liu et al., 2020).

The partition effect refers to the distribution of target pollutant molecules between the adjacent water layer on the surface of the MPs and the host solution. It mainly relies on van der Waals forces as a form of linear adsorption. The n value of the adsorption isotherm, corresponding to the adsorption of aliphatic and aromatic organic compounds by PE, is ~1, indicating strong linearity; therefore, these two types of organic compounds are adsorbed to the MPs via the partition effect (Hüffer and Hofmann, 2016). Guo and Wang (2019) applied the Freundlich model to confirm that the n value for the adsorption of antibiotics by PE and PS in freshwater systems is also ~1; thus, the isotherm is approximately linear. Therefore, the partition effect is also an important mechanism by which MPs adsorb antibiotics. Liu et al. (2019a) studied the adsorption of diethyl phthalate and dibutyl phthalate by PE, PVC, and PS and revealed that their adsorption isotherms were strongly linear, suggesting that such adsorption could be primarily attributed to the partition effect (Figure 4.4).

4.2.1.3   Hydrogen Bonding, Halogen Bonding, π–π Interactions and van der Waals Forces

Hydrogen bonding, halogen bonding, π–π interactions, and van der Waals forces between MPs and organic pollutants also affect their adsorption behavior (Zhou et al., 2014). Hydrogen bonding develops between MPs with halogen atoms (such as PVC) and organic compounds (such as PCBs and DDT) (Mato et al., 2001, Wu et al., 2019). In contrast, π–π interactions occur between PS and aromatic compounds (PAHs and PCBs) (Hüffer and Hofmann, 2016). Organic compounds only experience non-specific van der Waals interactions in hexadecane.

Guo et al. (2005) observed that polyolefin MPs form weak hydrogen bonds with organic pollutants containing benzene rings, strengthening their affinity. Liu et al. (2019b) analyzed the interactions between MPs and CIP before and after adsorption via Fourier-transform infrared spectrometry and observed peaks at approximately 3,500 cm^{-1} attributed to the hydrogen bonds between MPs and CIP, indicating that hydrogen bonds are one possible adsorption mechanism.

The halogen bond between the halogen atom on bisphenol and π electrons or hydroxyl groups of organic benzene rings promotes the adsorption of bisphenol by PVC. Additionally, π–π interactions also promote the adsorption of organic pollutants onto MPs (Ghaffar et al., 2015). For example, the π–π interaction is critical during the adsorption by PS of benzene ring-containing organic pollutants (Hüffer and Hofmann, 2016, Rochman et al., 2013). The π–π interactions also occur during the adsorption of PCBs and CIP by PS (Liu et al., 2019b). Liu et al. (2020) observed that π–π interactions represent one of the mechanisms by which PS adsorbs ATV and CIP; however, as PS ages, its benzene ring-containing molecules fall off, leading to the weakening of π–π interactions. Van der Waals forces are weak interactions between molecules that do not involve covalent or ionic bonding and typically form between aliphatic polymers (such as PE and PVC). Chen et al. (2019) used Fourier-transform infrared spectroscopy to investigate the adsorption of TnBP and

TCEP onto PE and observed no significant shift in the characteristic bands of TnBP and TCEP after adsorption onto PE when comparing their IR spectra. They suggested that weak van der Waals forces were the main adsorption forces between OPEs and PE (Figure 4.4).

4.2.2 FACTORS EFFECTING THE INTERACTION OF ORGANIC POLLUTANTS WITH MICROPLASTICS

The interaction process of organic pollutants with microplastics is mostly influenced by physicochemical properties of organic pollutants and microplastics. Besides, several environmental factors also effect the interaction process as indicated in Figure 4.5.

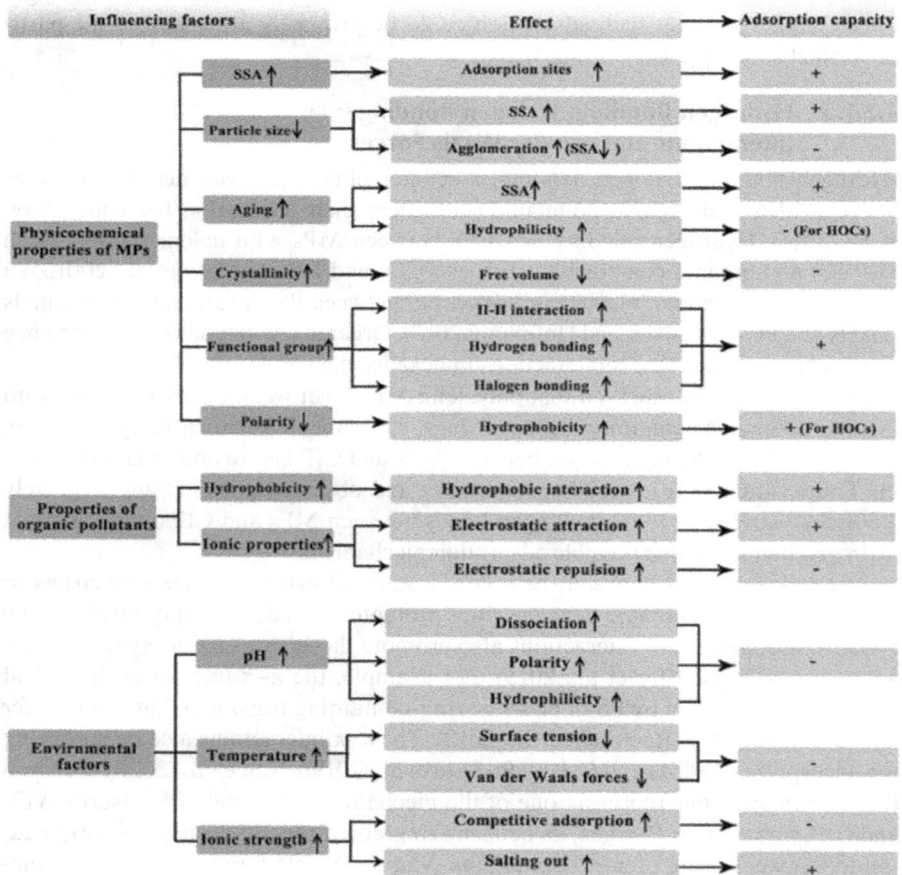

FIGURE 4.5 Factors affecting the adsorption capacity of MPs. (Adopted after permission from Fu et al., 2021.)

4.2.2.1   Effect of the Physicochemical Characteristics of Microplastics

Generally, the smaller the particle size and the larger the specific surface area (SSA) of MPs, the greater will be their number of adsorption sites and the amount of pollutants they can adsorb (Enders et al., 2015).

Zhang et al. (2019a) observed the adsorption behavior of 3,6-1,3,6,8-Tetrabromocarbazole (1,3,6,8-BCZ) and Dibromocarbazole (3,6-BCZ) and on polypropylene and the outcomes revealed that rates of adsorption for both compounds improved from 12%–28% to 11%–38%, respectively when the particle size of propylene reduce from 2–5 to 0.15–0.45 mm. Wang et al. (2019a) reported that the Log K_d values of PHE and nitrobenzene increased with decreasing particle size from 170 μm to 235 nm, which was attributed to the increase in SSA (from $0.4 \, m^2 g^{-1}$ in 170 μm PS to $27.6 \, m^2 g^{-1}$ in 235 nm PS). However, the particle size of MPs is not always inversely proportional to their SSA (Wang et al., 2019a). As a result, the Log K_d values of nano-polystyrene (50 nm) were less than the theoretical value of polystyrene (235 nm) (Enders et al., 2015).

Aging: In the natural environment, MPs undergo changes in their surface structure and oxygen-containing functional groups due to external environmental factors, including ultraviolet radiation, water (corrosion), and temperature (Mao et al., 2020, Sun et al., 2020). These factors also affect the mechanism by which MPs adsorb organic pollutants, as well as their adsorption capacities (Xu et al., 2020).

Scanning electron microscopy analysis revealed that, as MPs age, they become more oxidated and develop local microcracks on their surface. According to testing, the average pore diameter of weathered PS was 5.1 ± 0.2 nm. At the same time, the original PS was 39.3 ± 0.5 nm, which indicates that many small pores formed during weathering that increased the SSA of PS from 2.0 ± 0.1 to $7.9 \pm 0.2 \, m^2 g^{-1}$ (Sun et al., 2020).

Additionally, during the aging of MPs, some of their bonds, including C–H and C–C, are oxidized. The resultant oxygen-containing functional groups increase their hydrophilicity or strengthen the hydrogen bonds between MPs and organic pollutants, thereby improving their adsorption capacity for hydrophilic organic pollutants. Studies indicated that during photo-induced aging, characteristic peaks of –OH were observed at 3,446.17 and $3,437.60 \, cm^{-1}$ in the Fourier-transform infrared spectra of aged PS and PVC, respectively, but were not observed in pristine samples because the C–H bonds in the original MPs are broken, and hydroperoxides are generated. PVC exhibits a peak at $1,793.5 \, cm^{-1}$ due to the absorption of carbonyl moieties, which is the characteristic peak of carbonyl groups (C=O) and vinyl esters at 1,770–$1,800 \, cm^{-1}$ (Liu et al., 2018).

The crystalline region, which has regularly arranged molecular segments, is firmly held together and can easily condense. In contrast, the amorphous region has irregular molecular segments and is loose and soft, similar to a viscous liquid (Teuten et al., 2009). The amorphous region of MPs has a larger free volume, and organic pollutants have a greater affinity for this amorphous region than the crystalline region. So higher crystallinity of MPs decreases the chances of adsorption of organic pollutants.

The complexity, chain configuration, and glass transition temperature (T_g) of plastic polymers affect the crystallinity of MPs. The T_g, which represents the temperature at which the amorphous region of plastics changes from a rubber-like to a glass-like form, is an important factor affecting the adsorption capacity. Plastics with a T_g below the ambient temperature are referred to as rubber-like plastics, while those with a T_g above the ambient temperature are referred to as glass-like plastics. Pascall et al. (2005) reported that rubber-like PE and PP have higher PAH and PCB adsorption capacities than PET and PVC due to the larger free volume of their inner cavities.

MPs' functional groups and polarity determine their sorption behaviors toward organic pollutants. Hüffer and Hofmann (2016) studied the adsorption properties of PA, PE, PVC, and PS. They found that PS has the strongest adsorption capacity due to a strong π–π interaction between PS and toluene, facilitating high adsorption levels. Studies have also demonstrated that PA's AMX, TC, and CIP adsorption capacities exceed PS due to hydrogen bonds between amide groups (proton donor groups) in PA and carbonyl groups (proton acceptor groups) in AMX, TC, and CIP. Liu et al. (2019) reported that non-polar PE, PP, and PS are strongly hydrophobic, which increases their adsorption capacity for E2 beyond that of polar PVC, polycarbonate, polymethyl methacrylate, and strongly polar PA.

4.2.2.2   Effect of Properties of Organic Pollutants

Different organic pollutants have varying hydrophobicities, and they are adsorbed by various mechanisms and to varying degrees by different MPs. Generally, the hydrophobicity of organic pollutants is a key factor affecting the degree to which they are adsorbed. Organic pollutants with a large number of chlorine substituents are readily adsorbed by PS; similarly, charged MPs readily adsorb dissociable organic pollutants due to electrostatic action (Wang et al., 2019b).

4.2.2.3   Effect of Environmental Factors

Under different pH values, MPs and organic pollutants have different charges that affect adsorption. Increases in pH promote the dissociation of dissociable organic pollutants, resulting in negatively charged hydrophilic substances, which reduces the hydrophobic effect and triggers electrostatic repulsion between MPs and organic pollutants. Wang et al. (2015) reported that the PFOS adsorption capacities of PE and PS increase as the solution pH decreases. With a decrease in pH from 7 to 3, the level of PFOS adsorption onto PE and PS increases from 0.6 to 1.8 and 0 to 0.3 μg g^{-1}, respectively, because at pH 3.0–7.0, PFOS predominantly exists in its anionic form. At this pH the surfaces of PE and PS particles become protonated when the pH decreases that increasing PFOS anions adsorption. The 9-Nantes adsorption capacities of PP and PS decrease as the pH increases. When the pH exceeds 7, the negative potential of PP and PS and the polarity of 9-Nantes increase, resulting in strong electrostatic repulsion between the MPs and 9-Nantes (Zhang et al., 2020). Zhang et al. (2018a) reported that the oxytetracycline (OTC) adsorption capacity of aged PS is highest at pH 5. At pH < 5 or > 5, strong electrostatic repulsion occurs between PS and OTC, decreasing the adsorption capacity.

Temperature can significantly affect the adsorption capability of MPs. The promising reason for this behavior is that the adsorption process is an endothermic reaction; therefore, the spontaneity of the adsorption process may have intensified as the temperature raised. The adsorption of organic pollutants onto MPs at low temperatures was observed due to increased surface tension and reduced solubility of organic pollutants in water. The maximum adsorption capacities of PE and PVC for TnBP and TCEP at 15°C are 1,150 and 905 ng g⁻¹, respectively. As the temperature rose above 15°C, a decrease in the adsorption capacity was observed because of the more intense random molecular thermal motion of TnBP and TECP (Chen et al., 2019). Liu et al. (2018) also describe the similar behavior for adsorption of hexabromocyclododecane and tris-(2,3-dibromopropyl) isocyanurate on MPs as the temperature rises because of surface tension and van der Waals forces.

Solution ionic strength also plays a vital role in interacting with organic pollutants and microplastics. When MPs adsorb organic pollutants via the electrostatic mechanism, the salt ions and organic substances compete for adsorption sites on the MPs, thereby reducing the adsorption of organic pollutants. When the mechanism is related to hydrophobicity, the presence of salt ions induces the "salting out" effect, which lowers the solubility of the organic pollutants and promotes their hydrophobic interaction with the MPs. Zhang et al. (2018a) studied the influence of the ionic strengths of $NaCl$, $CaCl_2$, and Na_2SO_4 on the OTC adsorption by PE, and their results demonstrated that the adsorption of OTC by PE decreased with increasing ionic strength as Ca^{2+} and Na^+ compete with OTC for cation exchange sites on the surface of MPs. Additionally, they observed that the adsorption of OTC by MPs in the presence of $CaCl_2$ was stronger than NaCl or Na_2SO_4, indicating that ternary complexes formed between OTC, Ca^{2+}, and the functional groups on the surface of the MPs, thereby promoting adsorption. Similarly, the presence of Ca^{2+} and Na^+ can occupy the adsorption sites on the surfaces of MPs and decrease the adsorption of 9-NAnt by PP and PS.

Organic matters may affect the sorption of organic compounds by NPs/MPs through complex interactions. It was reported that the surface properties of MPs could be changed by the presence of dissolved organic matter (Chen et al., 2019). Xu et al. (2018) studied that fulvic acid had a significant effect on the sorption of TC (tetracycline) by three MPs. When fulvic acid was added to 20 mg C L⁻¹, the adsorbed concentration of TC on the MPs was less than 10 µg g⁻¹, which might be due to the higher affinity of TC with organic matter than with MPs. Zhang et al. (2018a) reported that with the increase of dissolved organic matter concentration, humic acid had a significant promotion effect on the sorption of OTC by beached MPs, possibly because humic acid acted as a bridge between the surface of the beached PS foam and the TC molecule.

Previous literature on adsorption kinetics of organic pollutants onto MPs indicated that adsorption is highly dependent on physicochemical properties of MPs, organic pollutants, and the surrounding environment as indicated in Table 4.2.

ACKNOWLEDGMENTS

This work was supported by a grant from the National Research Foundation of Korea (NRF) grant funded by the Korea government (MSIT) (No. NRF-2020R1A2C1013851).

TABLE 4.2

Adsorption Isotherm and Kinetics of organic pollutants on MPs

		Best Fit Model		
MPs	**Organic Pollutants / Adsorption Environment**	**Kinetic Models**	**Isotherm Models**	**Remarks**	**References**
PE, PS, PVC, PP	Fipronil/ Freshwater	Pseudo-second-order mode and Intraparticle diffusion models	Langmuir model	Intraparticle diffusion and film diffusion were rate-controlling step in the whole sorption process; monolayer coverage might be the predominant mechanism.	Gong et al. (2019)
PE, PS, PVC	Pyrene/ Freshwater	Pseudo-second-order mode and Intraparticle diffusion model	Langmuir model	Rate-limiting step was governed by both film diffusion and intraparticle diffusion; monolayer coverage might be the predominant mechanism.	Wang and Wang (2018)
	Ciprofloxacin (CIP)/ Freshwater	Pseudo-second-order mode	Freundlich model	The adsorption of CIP onto MPs was multilayer adsorption.	Liu et al. (2019)
PE	Pesticides: Carbendazim (CAR), Dipterex (DIP), Diflubenzuron (DIF), Malathion (MAL), Difenoconazole (DIFE)/ Freshwater	Pseudo-second-order model	Freundlich model	The adsorption process was a multilayer adsorption which was affected by physical and chemical adsorption. Besides, other rate-limiting steps were involved during adsorption process.	Wang et al. (2020)
PE	Imidacloprid, Buprofezin, Difenoconazole/ Freshwater	Pseudo-first-order model	Freundlich model	Physical adsorption was the leading force for the adsorption of pesticides on microplastics; adsorption processes were multilayer adsorption.	Li et al. (2021)
PS	Oxytetracycline/ Freshwater	Film diffusion model	Freundlich model	Both intraparticle diffusion and film diffusion were involved in the adsorption process; Oxytetracyline was adsorbed nonlinearly on MPs.	Zhang et al. (2018a)

(Continued)

TABLE 4.2 (Continued)
Adsorption Isotherm and Kinetics of organic pollutants on MPs

MPs	Organic Pollutants / Adsorption Environment	Best Fit Model		Remarks	References
PS	Triadimenol, Myclobutanil and Hexaconazole/ Ultrapure water	Pseudo-second-order model and Intraparticle diffusion models	Freundlich model	The adsorption process was nonlinear and easily occurred on the heterogeneous surface of the PS. Besides, subsequent pore filling may drive the adsorption.	Wang et al. (2020)
PP	Tris-(2,3-dibromopropyl) isocyanurate (TBC); Hexabromocyclododecanes/ Simulated seawater	Pseudo-first-order model	Langmuir model	The adsorption process was monolayer coverage, and the reaction rate was limited by the adsorbed sites.	Liu et al. (2018)
PP	3,6-Dibromocarbazole(3,6-BCZ); 1,3,6,8-Tetrabromocarbazole (1,3,6,8-BCZ)/ Simulated seawater	Pseudo-second-order mode	Langmuir model and Freundlich model	Monomolecular layer WAS formed in the initial stage of adsorption, and when monomolecular layer was saturated, multilayer adsorption occurred.	Zhang et al. (2019)

PE, polyethylene; PS, polystyrene; PVC, polyvinyl chloride; PA, polyamide; PET, polyethylene terephthalate; Pp; polypropylene

REFERENCES

Abdurahman, A., Cui, K., Wu, J., Li, S., Gao, R., Dai, J., Zeng, F. (2020) Adsorption of dissolved organic matter (DOM) on polystyrene microplastics in aquatic environments: kinetic, isotherm and site energy distribution analysis. *Ecotoxicology and Environmental Safety* 198: 110658.

Aghilinasrollahabadi, F., Salehi, M., Fujiwara, T. (2021) Investigate the influence of microplastics weathering on their heavy metals uptake in stormwater. *Journal of Hazardous Materials* 408, Article 124439.

Ashton, K., Holmes, L., Turner, A. (2010) Association of metals with plastic production pellets in the marine environment. *Marine Pollution Bulletin* 60(11): 2050–2055.

Brennecke, D., Duarte, B., Paiva, F., Caçador, I., Canning-Clode, J. (2016) Microplastics as vector for heavy metal contamination from the marine environment. *Estuarine, Coastal and Shelf Science* 178: 189–195.

Breivik, K., Sweetman, A., Pacyna, J. M., Jones, K. C. (2007) Towards a global historical emission inventory for selected PCB congeners – A mass balance approach. III. An update. *Science of the Total Environment* 377: 296–307.

Cao, Y., Zhao, M., Ma, X., Song, Y., Zuo, S., Li, H., Deng, W. (2021) A critical review on the interactions of microplastics with heavy metals: mechanism and their combined effect on organisms and humans. *Science of the Total Environment* 788: 147620, ISSN 0048-9697, doi: 10.1016/j.scitotenv.2021.147620.

Chen, S., Tan, Z., Qi, Y., Ouyang, C. (2019) Sorption of tri-n-butyl phosphate and tris (2-chloroethyl) phosphate on polyethylene and polyvinyl chloride microplastics in seawater. *Marine Pollution Bulletin* 149: 110490.

Crini, edited by Grégorio; Badot, Pierre-Marie. (2010) *Sorption Processes and Pollution: Conventional and Non-conventional Sorbents for Pollutant Removal from Wastewaters.* Besançon: Presses universitaires de Franche-Comté. p. 43. ISBN 978-2848673042.

Dong, Y., Gao, M., Song, Z., Qiu, W. (2020) As (III) adsorption onto different-sized polystyrene microplastic particles and its mechanism. *Chemosphere* 239: 124792.

Dong, Y. M., Gao, M. L., Song, Z. G., Qiu, W.W. (2019) Adsorption mechanism of As(III) on polytetrafluoroethylene particles of different size. *Environmental Pollution* 254, Article 112950.

Enders, K., Lenz, R., Stedmon, C. A., Nielsen, T. G. (2015) Abundance, size and polymer composition of marine microplastics $\geq 10\,\mu m$ in the Atlantic Ocean and their modelled vertical distribution. *Marine Pollution Bulletin* 100(1): 70–81.

Endo, S., Droge, S. T., Goss, K. U. (2011) Polyparameter linear free energy models for polyacrylate fiber– water partition coefficients to evaluate the efficiency of solid-phase microextraction. *Analytical Chemistry* 83(4): 1394–1400.

Fang, S., Yu, W., Li, C., Liu, Y., Qiu, J., Kong, F. (2019) Adsorption behavior of three triazole fungicides on polystyrene microplastics. *Science of the Total Environment* 691: 1119–1126.

Fred-Ahmadu, O. H., Bhagwat, G., Oluyoye, I., Benson, N.U., Ayejuyo, O. O., Palanisami, T. (2020) Interaction of chemical contaminants with microplastics: principles and perspectives. *Science of the Total Environment* 706: 135978.

Fu, Q., Tan, X., Ye, S., Ma, L., Gu, Y., Zhang, P., Tang, Y. (2020) Mechanism analysis of heavy metal lead captured by natural-aged microplastics. *Chemosphere* 270: 128624.

Fu, L., Li, J., Wang, G., Luan, Y., Dai, W. (2021) Adsorption behavior of organic pollutants on microplastics. *Ecotoxicology and Environmental Safety* 217: 112207, ISSN 0147-6513.

Fu, D., Zhang, Q., Fan, Z., Qi, H., Wang, Z., Peng, L. (2019) Aged microplastics polyvinyl chloride interact with copper and cause oxidative stress towards microalgae Chlorella vulgaris. *Aquatic Toxicology* 216, Article 105319.

Ghaffar, A., Ghosh, S., Li, F., Dong, X., Zhang, D., Wu, M., Pan, B. (2015) Effect of bio-char aging on surface characteristics and adsorption behavior of dialkyl phthalates. *Environmental Pollution* 206: 502–509.

Gong, W., Jiang, M., Han, P., Liang, G., Zhang, T., Liu G. (2019) Comparative analysis on the sorption kinetics and isotherms of fipronil on nondegradable and biodegradable microplastics. *Environmental Pollution* 254, Article 112927.

Guo, W., Guo, F., Wu, X., Tong, J., Wang, Z. (2005) Role of CH/pi, CH/O Weak Hydrogen Bonds in Constructing Inclusion Compounds of 2, 6-Bis (alpha-phenylbenzyl)-1, 5-naphthalenediol. *Acta Chimica Sinica-Chinese Edition* 63(16): 1525.

Guo, X., Hu, G., Fan, X., Jia, H. (2020) Sorption properties of cadmium on microplastics: the common practice experiment and a two-dimensional correlation spectroscopic study. *Ecotoxicology and Environmental Safety* 190: 110118.

Gao, F. L., Li, J. X., Sun, C.J., Zhang, L. T., Jiang, F. H., Cao, W., Zheng. L. (2019) Study on the capability and characteristics of heavy metals enriched on microplastics in marine environment. *Marine Pollution Bulletin* 144: 61–67.

Guo, X., Pang, J., Chen, S., Jia, H. (2018) Sorption properties of tylosin on four different microplastics. *Chemosphere* 209: 240–245.

Guo, X., Wang, J. (2019) Sorption of antibiotics onto aged microplastics in freshwater and seawater. *Marine Pollution Bulletin* 149: 110511.

Hodson, M. E., Duffus-Hodson, C.A., Clark, A., Prendergast-Miller, M.T., Thorpe. K.L. (2017) Plastic bag derived-microplastics as a vector for metal exposure in terrestrial invertebrates. *Science of the Total Environment* 51(8): 4714–4721.

Hansen, E., Nilsson, N. H., Lithner, D., Lassen, C. (2013) Hazardous substances in plastic materials. COWI in cooperation with Danish Technological Institute. On behalf of The Norwegian Climate and Pollution Agency, Oslo.

Holmes, L. A., Turner, A., Thompson, R. C. (2012) Adsorption of trace metals to plastic resin pellets in the marine environment. *Environmental Pollution* 160: 42–48.

Holmes, L. A., Turner, A., Thompson, R. C. (2014) Interactions between trace metals and plastic production pellets under estuarine conditions. *Marine Chemistry* 167: 25–32. doi: 10.1016/j.marchem.2014.06.001.

Hüffer, T., Hofmann, T. (2016) Sorption of non-polar organic compounds by micro-sized plastic particles in aqueous solution. *Environmental Pollution* 214: 194–201.

Langmuir, I. (1918) The adsorption of gases on plane surfaces of glass, mica and platinum. *Journal of the American Chemical society* 40(9): 1361–1403.

Li, Y., Li, M., Li, Z., Yang, L., Liu, X. (2019) Effects of particle size and solution chemistry on Triclosan sorption on polystyrene microplastic. *Chemosphere* 231: 308–314.

Lin, L. N., Gao, M. L., Song, Z. G., Mu. H.Y. (2020) Mitigating arsenic accumulation in rice (Oryza sativa L.) using Fe-Mn-La-impregnated biochar composites in arsenic-contaminated paddy soil. *Environmental Science and Pollution Research* 27(-4): 41446–41457.

Lin, Z., Hu, Y., Yuan, Y., Hu, B., Wang, B. (2021) Comparative analysis of kinetics and mechanisms for Pb (II) sorption onto three kinds of microplastics. *Ecotoxicology and Environmental Safety* 208: 111451.

Liu, F. F., Liu, G. Z., Zhu, Z. L., Wang, S. C., Zhao, F. F. (2019a) Interactions between microplastics and phthalate esters as affected by microplastics characteristics and solution chemistry. *Chemosphere* 214: 688–694.

Liu, J., Zhang, T., Tian, L., Liu, X., Qi, Z., Ma, Y., Chen, W. (2019b) Aging significantly affects mobility and contaminant-mobilizing ability of nanoplastics in saturated loamy sand. *Environmental Science & Technology* 53(10): 5805–5815.

Liu, X., Zheng, M., Wang, L., Ke, R., Lou, Y., Zhang, X., Zhang, Y. (2018) Sorption behaviors of tris-(2, 3-dibromopropyl) isocyanurate and hexabromocyclododecanes on polypropylene microplastics. *Marine Pollution Bulletin* 135: 581–586.

Liu, F. Nord, N. B., Bester, K. Vollertsen, J. (2020) Microplastics removal from treated wastewater by a biofilter. *Water* 12, Article 1085.

Li, X., Liang, R., Li, Y., Zhang, Y., Wang, Y., Li. K. (2021) Microplastics in inland freshwater environments with different regional functions: a case study on the Chengdu Plain. *Science of the Total Environment* 789, Article 147938.

Mao, R., Lang, M., Yu, X., Wu, R., Yang, X., Guo, X. (2020) Aging mechanism of microplastics with UV irradiation and its effects on the adsorption of heavy metals. *Journal of Hazardous Materials* 393: 122515.

Mato, Y., Isobe, T., Takada, H., Kanehiro, H., Ohtake, C., Kaminuma, T. (2001) Plastic resin pellets as a transport medium for toxic chemicals in the marine environment. *Environmental Science & Technology* 35(2): 318–324.

Mei, W., Chen, G., Bao, J., Song, M., Li, Y., Luo, C. (2020) Interactions between microplastics and organic compounds in aquatic environments: a mini review. *Science of the Total Environment* 736: 139472, ISSN 0048-9697, doi: 10.1016/j.scitotenv.2020.139472.

Nethaji, S., Sivasamy, A., Mandal, A. B. (2013) Adsorption isotherms, kinetics and mechanism for the adsorption of cationic and anionic dyes onto carbonaceous particles prepared from Juglans regia shell biomass. *International Journal of Environmental Science and Technology* 10(2): 231–242.

Öz, N., Kadizade, G., Yurtsever, M. (2019) Investigation of heavy metal adsorption on microplastics. *Applied Ecology Environmental Research* 17: 7310.

Pham, C. T., Kameda, T., Toriba, A., Hayakawa, K. (2013) Polycyclic aromatic hydrocarbons and nitropolycyclic aromatic hydrocarbons in particulates emitted by motorcycles. *Environmental Pollution* 183: 175–183.

Pascall, M. A., Zabik, M. E., Zabik, M. J., Hernandez, R. J., (2005) Uptake of polychlorinated biphenyls (PCBs) from an aqueous medium by polyethylene, polyvinyl chloride, and polystyrene films. *Journal of Agricultural and Food Chemistry* 53: 164–169.

Qi, K., Lu, N., Zhang, S., Wang, W., Wang, Z., Guan, J. (2021) Uptake of Pb (II) onto microplastic-associated biofilms in freshwater: adsorption and combined toxicity in comparison to natural solid substrates. *Journal of Hazardous Materials* 411: 125115.

Ryan, P. G., Moore, C. J, van Franeker, J. A., Moloney, C. L., (2009) Monitoring the abundance of plastic debris in the marine environment. *Philosophical Transactions of the Royal Society of London B* 364: 1999–2012

Rochman, C. M., Manzano, C., Hentschel, B. T., Simonich, S. L. M., Hoh, E. (2013) Polystyrene plastic: a source and sink for polycyclic aromatic hydrocarbons in the marine environment. *Environmental Science & Technology* 47(24): 13976–13984.

Sun, Y., Yuan, J., Zhou, T., Zhao, Y., Yu, F., Ma, J. (2020) Laboratory simulation of microplastics weathering and its adsorption behaviors in an aqueous environment: a systematic review. *Environmental Pollution* 265: 114864.

Tang, C.-C., Chen, H.-I., Brimblecombe, P., Lee, C.-L. (2018) Textural, surface and chemical properties of polyvinyl chloride particles degraded in a simulated environment. *Marine Pollution Bulletin* 133: 392–401.

Tang, N., Liu, X., Xing, W. (2020) Microplastics in wastewater treatment plants of Wuhan, Central China: Abundance, removal, and potential source in household wastewater. *Science of the Total Environment* 745, Article 141026.

Teuten, E. L., Saquing, J. M., Knappe, D. R., Barlaz, M. A., Jonsson, S., Björn, A., Takada, H. (2009) Transport and release of chemicals from plastics to the environment and to wildlife. *Philosophical Transactions of the Royal Society B: Biological Sciences* 364(-1526): 2027–2045.

Titmus, A. J., Hyrenbach, K. D. (2011) Habitat associations of floating debris and marine birds in the North East Pacific Ocean at coarse and meso spatial scales. *Marine Pollution Bulletin* 62: 2496–2506.

Tu, C., Chen, T., Zhou, Q., Liu, Y., Wei, J., Waniek, J. J., Luo, Y. (2020) Biofilm formation and its influences on the properties of microplastics as affected by exposure time and depth in the seawater. *Science of the Total Environment* 734: 139237.

Verla, A. W., Enyoh, C. E., Verla, E. N. et al. (2019) Microplastic–toxic chemical interaction: a review study on quantified levels, mechanism and implication. *SN Applied Sciences* 1: 1400. doi: 10.1007/s42452-019-1352-0.

Wang, F., Shih, K. M., Li, X. Y. (2015) The partition behavior of perfluorooctanesulfonate (PFOS) and perfluorooctanesulfonamide (FOSA) on microplastics. *Chemosphere* 119: 841–847.

Wang, J., Liu, X., Liu, G., Zhang, Z., Wu, H., Cui, B., Zhang, W. (2019a) Size effect of polystyrene microplastics on sorption of phenanthrene and nitrobenzene. *Ecotoxicology and Environmental Safety* 173: 331–338.

Wang, Y., Li, M., Yu, H., Ma, G., Wei, X. (2019b) Research progress on the adsorption and desorption between microplastics and environmental organic pollutants. *Asian Journal of Ecotoxicology* 14: 23–30.

Wang, W., Wang, J. (2018) Comparative evaluation of sorption kinetics and isotherms of pyrene onto microplastics. *Chemosphere* 193: 567–573.

Wang, Q., Zhang, Y., Wang, J. X., Wang, Y., Meng, G., Chen. Y. (2020) The adsorption behavior of metals in aqueous solution by microplastics effected by UV radiation. *Journal of Environmental Sciences* 87: 272–280.

Wu, P., Cai, Z., Jin, H., Tang, Y. (2019) Adsorption mechanisms of five bisphenol analogues on PVC microplastics. *Science of the Total Environment* 650: 671–678.

Xu, B., Liu, F., Brookes, P. C., Xu, J. (2018) Microplastics play a minor role in tetracycline sorption in the presence of dissolved organic matter. *Environmental Pollution* 240: 87–94, ISSN 0269-7491.

Xu, P. C., Guo, J., Ma, D., Ge, W., Zhou, Z. F., Chai, C. (2020) Sorption of polybrominated diphenyl ethers by virgin and aged microplastics. *Huan Jing ke Xue Huanjing Kexue* 41(3): 1329–1337.

Zhang, H., Wang, J., Zhou, B., Zhou, Y., Dai, Z., Zhou, Q., Luo, Y. (2018a) Enhanced adsorption of oxytetracycline to weathered microplastic polystyrene: kinetics, isotherms and influencing factors. *Environmental Pollution* 243: 1550–1557.

Zhang, J., Chen, H., He, H., Cheng, X., Ma, T., Hu, J., Zhang, L. (2020) Adsorption behavior and mechanism of 9-Nitroanthracene on typical microplastics in aqueous solutions. *Chemosphere* 245: 125628.

Zhang, Q., Yang, C., Lin, S., Sun, H., Shen, Y., Tan, H., Lin. C. (2012) Determination of n-octanol /water partition coefficient of melamine. *Guizhou Academy of Science* 30: 60–62.

Zhang, X., Zheng, M., Wang, L., Lou, Y., Shi, L., Jiang, S. (2018b) Sorption of three synthetic musks by microplastics. *Marine Pollution Bulletin* 126: 606–609.

Zhang, P., Yan, Z., Lu, G., Ji, Y. (2019) Single and combined effects of microplastics and roxithromycin on Daphnia magna. *Environmental Science and Pollution Research* 26(17): 17010–17020.

Zou, J., Liu, X., Zhang, D., Yuan, X. 2020. Adsorption of three bivalent metals by four chemical distinct microplastics. *Chemosphere*, 248:126064.

Zhou, X., Wei, J., Liu, K., Liu, N., Zhou, B. (2014) Adsorption of bisphenol A based on synergy between hydrogen bonding and hydrophobic interaction. *Langmuir* 30(46): 13861–13868.

Zhou, Y., Liu, X., Wang, J. (2019) Characterization of microplastics and the association of heavy metals with microplastics in suburban soil of central China. *Science of the Total Environment* 694, Article 133798.

Zon, N. F., Iskendar, A., Azman, S., Sarijan, S., Ismail, R. (2018) Sorptive behaviour of chromium on polyethylene microbeads in artificial seawater. *In MATEC Web of Conferences EDP Sciences* 250: 06001.

5 Impacts of Microplastics and Nanoplastics on Biota

Sadia Ilyas and Hyunjung Kim
Hanyang University

Gukhwa Hwang
Jeonbuk National University

CONTENTS

Microplastics and nanoplastics are found in almost all aquatic and terrestrial environments making their potentially pernicious effects a global problem (Claessens et al., 2013). The difference in type, shape, and density causes these plastics to disperse diversely in different compartments of the aquatic and terrestrial environment (water surface, water column, sediment, and land) and influence their availability to organisms at different trophic levels and/or occupying different habitats (Betts, 2008; Cole et al., 2011). For example, pelagic organisms such as phytoplankton and small crustaceans (e.g., zooplankton) (Desforges et al., 2015) are more likely to encounter less dense, floating microplastics and nanoplastics while benthic organisms, including amphipods (Thompson et al., 2004), polychaete worms (Mathalon and Hill, 2014), tubifex worms (Hurley et al., 2018), mollusks (Brillant and MacDonald, 2002; Browne et al., 2007) and echinoderms (Hart, 1991; Graham and Thompson, 2009) are more likely to encounter microplastics (MPs) that are denser than water. Both benthic (de Sá et al., 2015) and pelagic (Rummel et al., 2016) fish may ingest MPs directly or indirectly (i.e., consume them in prey). Birds (Herzke et al., 2016) and mammals (Fossi et al., 2012) feeding on aquatic organisms or living in aquatic environments are also known to ingest MPs. Furthermore, it is also observed that plant and soil characteristics are altered after interaction with various microplastics (de Souza Machado et al., 2019). Potential negative effects of microplastics in the human body, focusing on pathways of exposure and toxicity, were observed. Exposure may occur by ingestion, inhalation, and dermal contact due to the presence of microplastics in products, foodstuff, and air. In all biological systems, microplastics and nanoplastics exposure may cause particle toxicity, oxidative stress, inflammatory lesions, and increased uptake or translocation. The inability of the immune system to remove synthetic particles may lead to chronic inflammation and increase the risk of neoplasia. Furthermore, microplastics may release their constituents, adsorbed contaminants, and pathogenic organisms. Nonetheless, knowledge of microplastic toxicity is still limited and largely influenced by exposure concentration, particle properties, adsorbed contaminants, tissues involved, and individual susceptibility, which requires further research (Prata et al., 2020).

5.1   IMPACT OF MICROPLASTICS AND NANOPLASTICS ON PLANT AND SOIL CHARACTERISTICS

Microplastics can affect the biophysical properties of the soil. However, little is known about the cascade of events in fundamental levels of terrestrial ecosystems, i.e., starting with the changes in soil abiotic properties and propagating across the various components of soil–plant interactions, including soil microbial communities and plant traits.

de Souza Machado et al. (2019) investigated the effects of six different types of MPs, polyester (PES) fibers, polyamide beads, polyethylene high density (PEHD), polyester terephthalate (PET), polypropylene (PP), and polystyrene (PS), on a broad suite of proxies for soil health and performance of spring onion (*Allium fistulosum*). Microplastic addition resulted in alteration of physical characteristics of soil (Figure 5.1a–c) with consequences for water dynamics and microbial activity (Figure 5.1d–f).

Soil bulk density was decreased by PEHD, PES, PET, PP, and PS, probability >97.5% (Figure 5.1a), while soil density in the rhizosphere was increased (probability >97.5%). Except for PP, an interactive effect of microplastic with plants was observed with a probability of >75.0%. Concomitant decreases in water stable aggregates were significant for PA and PES (probability >97.5%) and for PS (probability >75.0% (Figure 5.1b). Rhizosphere presented higher water-stable aggregates (probability

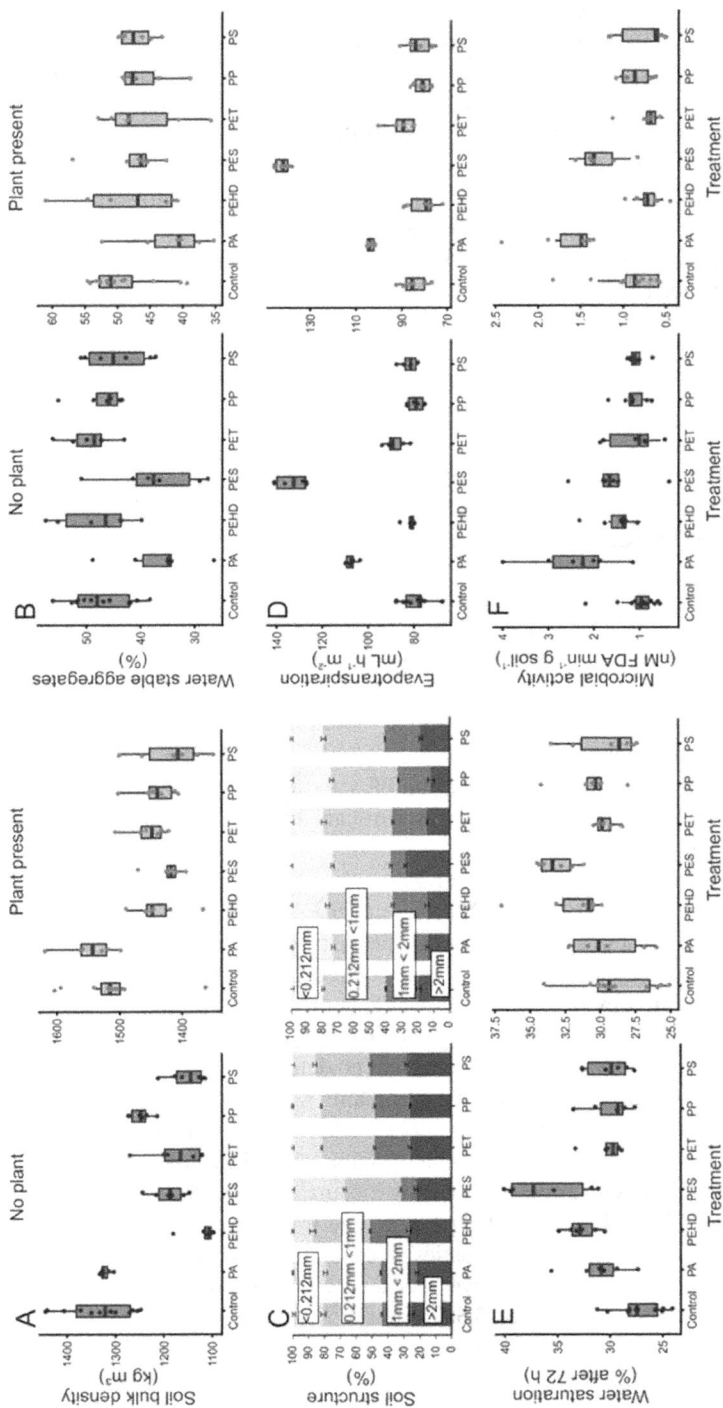

FIGURE 5.1 (a–f) Effects of microplastics on soil environment and function. (Retrieved from de Souza Machado et al., 2019 after permission from ACS through https://pubs.acs.org/doi/full/10.1021/acs.est.9b01339.)

>75.0%) and interacted with soils treated with PA, PES, PET, and PP (probability >75.0%). The soil structure was affected by all MPs, but the intensity of effects was highly dependent on the type of microplastics (Figure 5.1c), plant presence, and aggregate size fraction (probabilities >75.0%). Evapotranspiration was increased by ~35% by PA and ~50% by PES (probability >97.5%), and smaller increases were associated with PEHD, PET, and PS with probability >75.0%. Spring onions increased the evapotranspiration with probability >97.5% (Figure 5.1d), and most plastics interacted with the plants to either increase (e.g., PES) or decrease (e.g., PA) evapotranspiration (probability >97.5%). Increases in evaporation were smaller than increases in water holding capacity. Therefore, the water availability was generally higher in soils treated with microplastics with probability >97.5% (Figure 5.1e), which was attenuated by plants (probability >97.5%). In turn, the general microbial metabolic activity was increased by PA, PEHD, and PES with probabilities >97.5% (Figure 5.1f) and decreased by interactive effects of plants and PA, PEHD, and PET (probabilities >75.0%).

PES and PS triggered significant increases in root biomass (probability >97.5%), while a weaker effect was observed in plants exposed to PEHD, PET, and PP (probability >75.0%) as indicated in Figure 5.2a. PA decreased the ratio between root and dry leaf biomass (Figure 5.2b) (probability >97.5%). At the same time, the exposure to PES, PET, and PP significantly increased this ratio (probability >75.0%) as well as exposure to PEHD and PS (probability >97.5%). Moreover, all tested microplastics significantly increased total root length with probabilities of >75.0%, as in Figure 5.2c and decreased root average diameter as can be seen in Figure 5.2d. With increased biomass of longer and finer roots, the total root area was increased by all microplastics with probabilities of >75.0% as in Figure 5.2e. PA caused a decrease in root tissue density (probability >97.5%); PES and PS triggered an increase of such response (probability >75.0%), and no significant effect was observed for PEHD, PET, and PP (Figure 5.2f).

Root symbioses were also affected by microplastic treatments. PES increased ~8-fold the root colonization by AMF (Figure 5.2g), while PP caused an ~1.4-fold increase, and PET caused a reduction of ~50% in root colonization. PES triggered the strongest effect on the interaction of roots and the surrounding microbial communities as measured by increased mycorrhizal coils and non-AMF structures. With the exception of PP causing a small increase in colonization by coils and PA decreasing non-AMF structures, other treatments had no detectable effects. The dry biomass of onion bulbs was decreased in PA-treated plants while it nearly doubled after PES exposure (Figure 5.3a). In fact, all microplastic treatments were significantly different from control regarding the dry weight of onion bulbs. Likewise, the water content of onion bulbs increased 2-fold under PA exposure as indicated in Figure 5.3b and decreased with PES, PET, and PP exposure. The water content of the aboveground tissue was less sensitive to microplastics, with significant increases observed only for PA and PES (Figure 5.3c). However, PA increased leaf nitrogen content, and PES significantly decreased it (Figure 5.3d). Thus, PA significantly decreased the C–N ratio, and PES increased it (Figure 5.3e). Total biomass was increased by PA and PES (Figure 5.3f). In the first case, the effect was driven by increases in the aboveground leaf, while for the latter, there were increases in the belowground bulb. It is worth

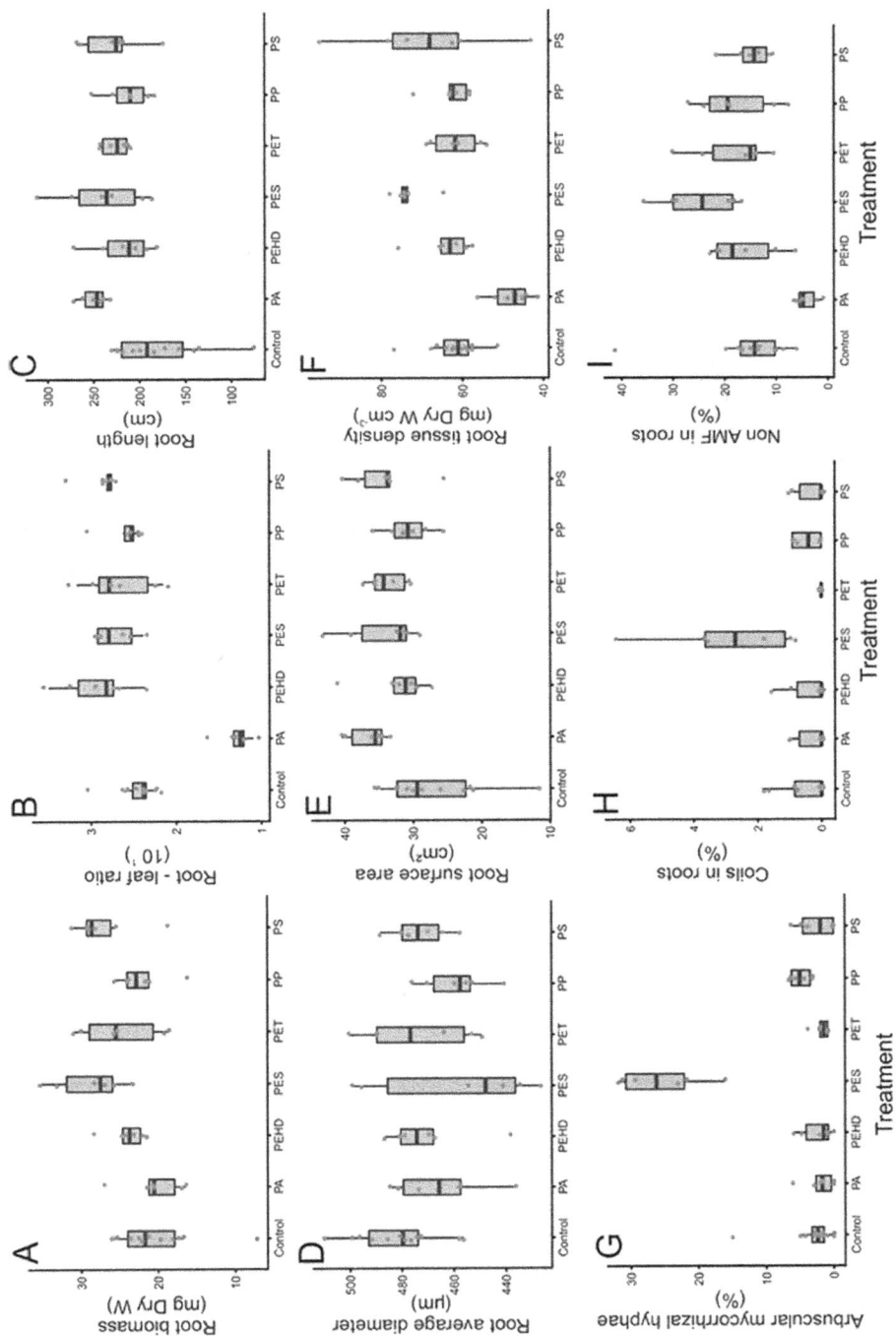

FIGURE 5.2 (a–i) Effects of microplastics on root traits. (Retrieved from de Souza Machado et al., 2019 after permission from ACS through https://pubs.acs.org/doi/full/10.1021/acs.est.9b01339.)

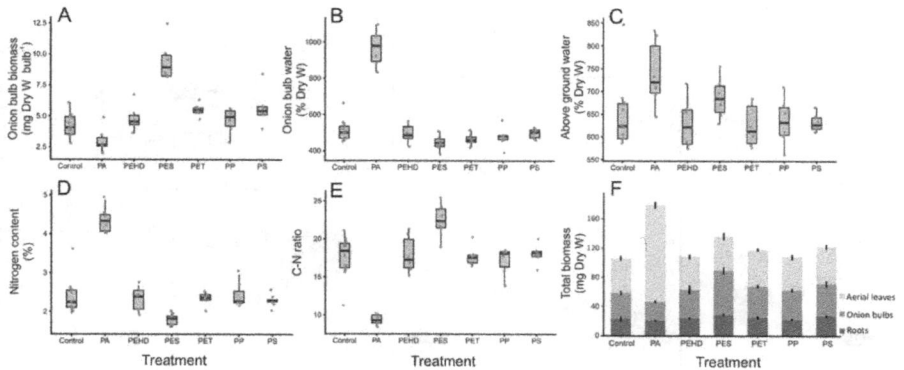

FIGURE 5.3   (a–f) Effects of microplastics on proxies of general plant fitness. (Retrieved from de Souza Machado et al., 2019 after permission from ACS through https://pubs.acs.org/doi/full/10.1021/acs.est.9b01339.)

mentioning that PA contains nitrogen in its composition, which might be accountable for the observed effects. Further increases in total biomass were observed for PET and PS. None of the microplastic treatments significantly decreased total biomass (de Souza Machado et al., 2019).

5.2   IMPACT OF MICROPLASTICS AND NANOPLASTICS ON INVERTEBRATES

The Invertebrates unit explores six groups of invertebrates include poriferans (sponges), cnidarians (such as sea jellies and corals), echinoderms (such as sea urchins and sea stars), mollusks (such as octopuses, snails, and clams), annelids (worms), and arthropods (such as insects, spiders, and lobsters). MPs alone and with other contaminants can have some impact on these organisms. Figures 5.4 depict the individual and combined (along with contaminants) ecotoxicological effect of microplastics on some invertebrates.

5.2.1   IMPACT OF MICROPLASTICS AND NANOPLASTICS ON PORIFERANS

Sponges are an important component of temperate benthic ecosystems, providing a range of important functional roles. Sponges can adapt to many environments by exploiting various food sources, from dissolved organic matter to small crustaceans. Regardless of this, sponges feed primarily on picoplankton and can retain up to 99% of these from seawater. To examine the effects of MP on sponge respiration, two temperate sponge species (*Tethya bergquistae* and *Crella incrustans*) were exposed to two different-sized plastic particles (1 and 6 μm) at two different concentrations (200,000 and 400,000 beads per mL). Results indicate that sponges are resilient to MP pollution. The only significant result was the effect of MP size on the respiration rates on *Tethya bergquistae* ($P = 0.001$), but there were no other significant main effects or interactions (http://researcharchive.vuw.ac.nz/bitstream/handle/10063/6749/thesis_access.pdf?sequence=1).

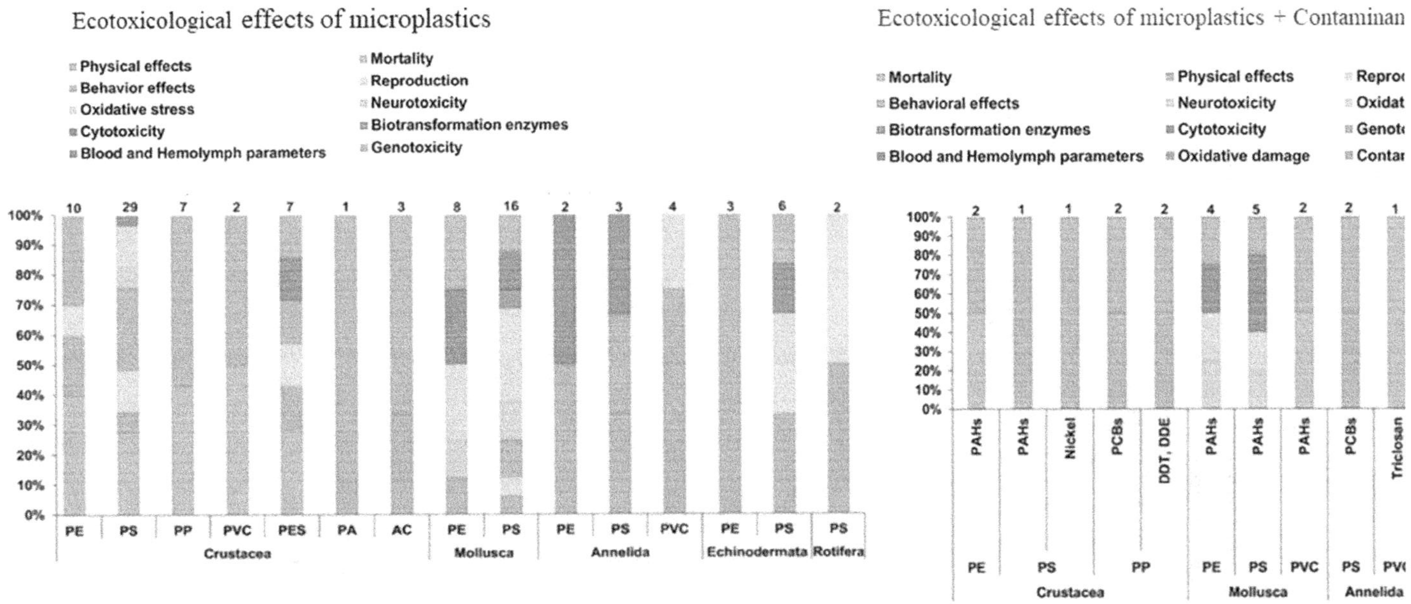

FIGURE 5.4 Individual and combined (along with contaminants) ecotoxicological effect of microplastics on some invertebrates. (Modified from Carlos de Sá et al., 2018 under a Creative Commons License copyright Elsevier.)

5.2.2 IMPACT OF MICROPLASTICS AND NANOPLASTICS ON CNIDARIANS

Venâncio et al (2021) indicated the impact of microplastics and nanoplastics on freshwater cnidarian *Hydra viridissima*. *Hydra viridissima* were exposed to polymethylmethacrylate nanoplastics (PMMA-NPLs) for 96 h. High concentration (< 40 mg L^{-1}) of PMMA-NPLs can impair the survival of *H. viridissima*, with an estimated 96 h-LC50 of 84.0 mg PMMA-NPLs/L. Several morphological alterations were detected at concentrations below 40 PMMA-NPLs mg/L, namely partial or total loss of tentacles, which did not induce significant alterations in the feeding rates. Morphological alterations not previously reported in the literature were also found after the 96 h exposure, such as double or elbow-like tentacles. Exposure to 40 mg PMMA-NPLs/L significantly impacted hydra regeneration, with organisms exposed to PMMA-NPLs presenting significantly slower regeneration rates than controls but with no impacts on the feeding rates. Overall, this work highlights the need to assess the effects of NPLs in freshwater biota. *Hydra viridissima* species was sensitive to a wide range of endpoints showing its value as a biological model to study the effects of small plastic particles.

5.2.3 IMPACT OF MICROPLASTICS AND NANOPLASTICS ON ANNELIDS

Daly and Wania (2005) conducted research to compare the difference in biological effects on the lugworm (*Arenicola marina*) between polyvinyl chloride (PVC) and PEHD (average size 130.6 ± 12.9 and 102.6 ± 10.3, respectively) and one biodegradable plastic (polylactic acid; average size 235.7 ± 14.8 μm), it was found that the lugworm produced a reduced number of casts in the sediment that was contaminated with microplastics. Furthermore, by measuring the oxygen consumption, it was established that the metabolic rate of the lugworm was higher when exposed to the microplastics, indicating a stress response. The strongest effect was observed with PVC. The toxicity of PVC microplastics to the lugworm in the environment was hypothesized to result from the leaching of intrinsic plasticizers and vinyl chloride monomers or adsorbed persistent organic pollutants (POPs). Browne et al. (2013) documented increased mortality of the annelid worm *Arenicola marina* exposed to PVC MPs and the antibiotic triclosan.

5.2.4 IMPACT OF MICROPLASTICS AND NANOPLASTICS ON ECHINODERMS

Della Torre et al. (2014) described the effects of PS MPs (6–48 h; 0.04–0.05 μm; 1–50 mg L^{-1}) on gene expression in the echinoderm *Paracentrotus lividus*, including an up-regulation of the *Abcb1* gene responsible for protection and multi-drug resistance (Shipp and Hamdoun, 2012).

Norén and Naustvoll (2010) investigated the effects of virgin microplastics versus beach recovered microplastics on the development of green sea urchin (*Lytechinus variegatus*) embryos. Virgin microplastics exhibited the greatest toxicity due to the leaching of intrinsic additives. Furthermore, the degree of toxicity of the beach recovered microplastics was highly variable and was postulated to result from differences in the levels of sorbed contaminants.

5.2.5 Impact of Microplastics and Nanoplastics on Arthropods

Bergami et al. (2016) studied the interactions of 40 nm anionic carboxylated and 50 nm cationic amino polystyrene nanoplastics with brine shrimp (*Artemia franciscana*) larvae. They found that after 48 h of exposure to both types of nanoplastics at a concentration of 5–100 μg mL^{-1}, a large accumulation of nanoplastics was discovered in the central cavity of the brine shrimp larvae digestive tract and subsequent excretion was limited. Cole et al. (2015) studied small crustaceans and observed a decrease in survival and fecundity of the marine copepods *Calanus helgolandicus* (24 and 216 h; 20 μm; 6.5–7.5 × 104 particles L^{-1}) and *Tigriopus japonicas* (two-generation test; 0.05–0.5 μm; 0.125–25 mg L^{-1}) when exposed to PS MPs. Gambardella et al. (2017) and Jeong et al. (2016) demonstrated some alteration of enzymes in the small crustaceans Artemia franciscana (48 h; 0.1 μm; 0.001–10 mg L^{-1}) and *Paracyclopina nana* (24 h; 0.05–6 μm; 0.1–20 mg L^{-1}). Avio et al. (2015) documented similar enzyme alteration effects following PS and PE microplastics exposure in the marine *mollusc M. galloprovincialis* (168 h; <100 μm; 2 × 10^3 mg L^{-1}).

Effects of PP MPs on *Hyalella Azteca* (240 and 1,008 h; 20–75 μm; 0–9 × 104 particles L^{-1}) have been reported by Au et al. (2015), who demonstrated higher toxicity for PP than PE MPs. They reported a LC$_{50}$ = 7.14 × 104 particles L^{-1} for PP MPs compared to LC$_{50}$ = 4.64 × 107 particles L^{-1} for PE MPs.

5.2.6 Impact of Microplastics and Nanoplastics on Mollusk

Mollusks are rich in nutrition, have high economic value, and are easy to breed. They are a highly regarded aquaculture resource, especially bivalves, typical filter feeders, constantly filtering out microbes and organic matter from the surrounding water (Xu et al., 2017). As a result, it is critical to monitor MPs in mollusks. *Mytilus galloprovincialis* (168 h; <100 μm; 2 × 103 mg L^{-1}). Additional ecotoxicological effects of exposure to PS MPs were observed in two mollusk species. A study using *Scrobicularia plana* (Ribeiro et al., 2017) reported increased neurotoxicity and genotoxicity. A 25% increase in energy consumption was reported after ingestion of PS MPs by *Mytilus edulis* (336 h; <100 μm; 1.1 × 105 particles L^{-1}) (Van Cauwenberghe and Janssen, 2014), probably associated with an effort to digest inert material and maintain physiological homeostasis (von Moos et al., 2012). In *Mytilus edulis*, the transition of MP from the gut to the hemolymph was observed to continue for >48 days after 3 days of exposure (Browne et al., 2007).

5.3 IMPACT OF MICROPLASTICS AND NANOPLASTICS ON VERTEBRATES

Vertebrates are animals that have a backbone inside their body. Although outnumbered in species and biomass, they have ecological importance and have a diverse range in habitat utilization due to the vast differences in ecomorphological characteristics and the diversity in different taxa associated with marine ecosystems. This group can be roughly divided into seven different amphibians (e.g., frogs), reptiles (e.g., turtles), birds, fish, and mammals. Microplastics and nanoplastics have an impact on each group of vertebrates. Figures 5.5 depicts the individual and combined (along with contaminants) ecotoxicological effect of microplastics on some vertebrates.

5.3.1 Impact of Microplastics and Nanoplastics on Amphibians

A study by da Costa (2020) assesses the toxicological potential of PE MPs in *Physalaemus cuvieri* tadpoles. According to the results, tadpoles' exposure to MP PE at concentration of 60 mg L^{-1} for 7 days led to mutagenic effects, which were evidenced by the increased number of abnormalities observed in nuclear erythrocytes. The small size of erythrocytes and their nuclei area, perimeter, width, length, and radius, as well as the lower nucleus/cytoplasm ratio observed in tadpoles exposed to PE MPs, confirmed its cytotoxicity. External morphological changes observed in the animal models included a reduced ratio between total length and mouth-cloaca distance, caudal length, ocular area, and mouth area. PE MPs increased the number of melanophores in the skin and pigmentation rate in the assessed areas. Finally, PE MPs were found in gills, gastrointestinal tract, liver, muscle tissues of the tail, and blood.

5.3.2 Impact of Microplastics and Nanoplastics on Reptiles

Studies of loggerhead turtles (*Caretta caretta*) and green turtles (*Chelonia mydas*) have indicated that the consumption of plastic litter by these species results in nutritional deficiencies since plastic material cannot be digested and has no nutritional value (Desforges et al., 2014; Erdman et al., 2012).

Sala et al. (2021) indicated the presence of organophosphate esters (OPEs) in loggerhead turtles (*Caretta caretta*). OPEs were detected in all the samples, ranging from 6.18 to 100 ng g^{-1} wet weight. Levels were higher in the turtles from the Balearic Islands, but OPE profiles did not differ regionally. In turtles, OPE levels were in the same order of magnitude as other legacy pollutants (PCBs and DDTs).

Samples corresponding to the main turtle prey (jellyfish, squid, and sardine), as well as different types of marine plastic debris, were analyzed, also showing the presence of OPEs at concentration levels between 5.21 and 90.5 ng g^{-1} wet weight and 10.9–868 ng g^{-1}, respectively. Some differences were observed in turtles' OPE profiles, prey, and plastic debris. Moreover, it seems that turtles may have been contaminated by both prey and plastic debris. OPE biomagnification has been evaluated through data in predators and preys, and some compounds (TEP, DCP, 2IPPDPP, 4IPPDPP, and T2IPPP) seem to have biomagnification potential.

5.3.3 Impact of Microplastics and Nanoplastics on Fish

Fish may ingest microplastics directly or indirectly and the consumption of microplastics by fish can interfere with biological processes, such as the inhibition of gastrointestinal function and causing blockages and inducing feeding impairment (Mariussen, 2012; Xu et al., 2012; Plastics Europe, 2015). Furthermore, fish are often bio-accumulated with waterborne pollutants to various degrees. For example, a study

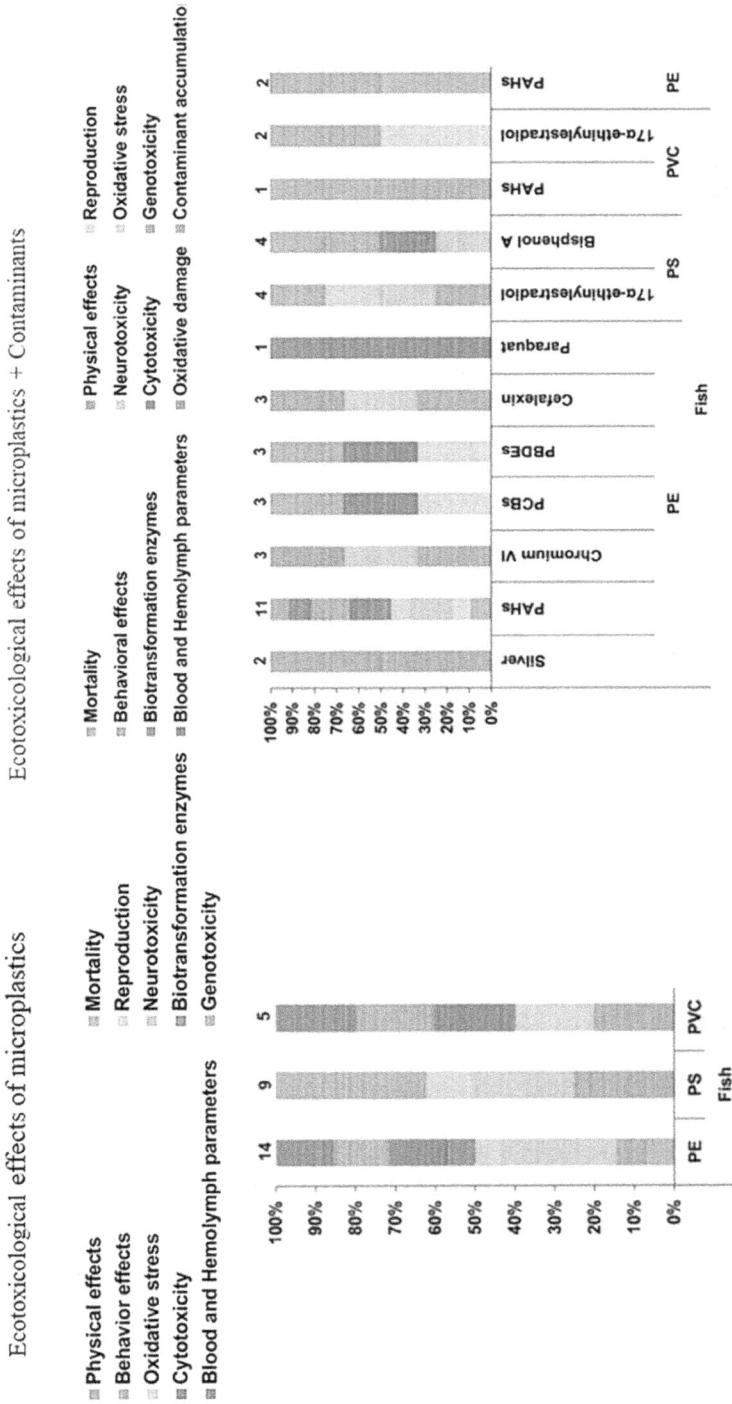

FIGURE 5.5 Individual and combined (along with contaminants) ecotoxicological effect of microplastics on some vertebrates. (Modified from Carlos de Sá et al., 2018 under a Creative Commons License copyright Elsevier.)

found high levels of polychlorinated biphenyls (PCBs) of 1,066–3,112 ng g⁻¹ in fish liver collected from an abandoned metropolitan dumping area for sewage sediment in Garroch Head, a Scottish coastal region in Argyll and Bute. A further study found more than 500 ng g⁻¹ concentrations of PCBs in North Atlantic fish, collected off the West coast of Scotland (Wei et al., 2012). An examination of fish products sold in the Canadian retail market in 2002 found total concentrations of PCBs to be on average 12.9 ng g⁻¹ (net weight) in the salmon (Reisser et al., 2013; Douabul et al., 1997; EU, 2008). In a study assessing the levels of perfluoroalkylated substances in wild eels in Comacchio Lagoon and Po River (located on the Northwest Adriatic coast), it was found that concentrations were in the ranges <0.4–92.77 ng g⁻¹ w.w. for perfluorooctanoic acid and <0.4–6.28 ng g⁻¹ w.w. for perfluorooctane sulfonic acid. The highest accumulations were found in blood, while the lowest was in muscle tissue. Furthermore, the researchers reported that lesions in the liver of the eels were observed, and perfluoroalkylated substances levels were found to be on par or lower than levels observed in other European fish. A study of fish collected from seas around China found concentrations of polybrominated diphenyl ethers (PBDEs) in the fish tissues to be 0.3–700 ng g⁻¹ (Löder and Gerdts, 2015).

Incidentally, fish are highly mobile creatures, and many travel great distances by their very nature. Thus, if fish are ingesting plastic, they may be unwittingly serving as a vessel for the long-range distribution of microplastics and POPs. For example, large abundances of microplastics have been reported in deep ocean sediments, and a study has identified plastic ingestion by deep-sea fish. Mostly, deep ocean fish are more prone to bio-accumulation of high levels of pollutants, such as PCBs and PBDEs pollutants, on account of their typical extended lifespans and their propensity to consume organisms situated further up the food web, in contrast to their epipelagic counterparts (Wesch et al., 2016; Wu et al., 2014; Webster et al., 2011).

Probably, the capability of microplastics to become contaminated with hydrophobic pollutants via sorptive processes facilitates the transfer of these hydrophobic pollutants to fish if the contaminated microplastics are ingested, thereby inducing toxicological effects. For example, it was determined that when European seabass (*Dicentrarchus labrax*) were fed <0.3 mm of PVC microplastics, which had been exposed to seawater in Milazzo Harbour, Italy for 3 months to adsorb contaminants. Upon examination, it was discovered that there were significant pathological alterations to the fish's distal part of the intestine. Exposure to >0.3 mm virgin PVC microplastics resulted in a similar effect (Lenz et al., 2015; Ivar do Sul and Costa, 2014). However, the polluted PVC microplastics exhibited more damage, and it was evident that PVC microplastics could completely compromise intestinal function. Furthermore, as exposure times of the fish to PVC microplastics were extended, the pathological alterations increased from moderate to severe. In another study, fish were exposed for 96h to pyrene (20 and 200 µgL⁻¹) in the absence and presence of microplastics (0, 18.4 and 184 µgL⁻¹). Mortality, bile pyrene metabolites, and biomarkers involved in neurotransmission, aerobic energy production, biotransformation and oxidative stress were quantified. Microplastics delayed pyrene-induced fish mortality and increased the concentration of bile pyrene metabolites. Microplastics, alone or in combination with pyrene, significantly reduced acetylcholinesterase (AChE) activity, an effect also observed for pyrene alone. The mixture also decreased isocitrate dehydrogenase (IDH) activity. No significant effects were found for glutathione S-transferase activity or lipid peroxidation (Oliveira et al., 2013).

Additionally, the interactions of microplastics with heavy metals and their effect on these fish have been studied. For example, in a study of the effects of a mixture of chromium (VI) and 1–5 μm polyethylene mini-microplastics on early juveniles of the common goby (*P. microps*), obtained from the estuaries of the Lima and Minho rivers in the Northwest Iberian Peninsula, it was observed that toxicological interactions between the microplastics and chromium (VI) occurred. The researchers established that in the fish collected from the Lima River, chromium (VI) concentrations of more than 3.9 mg L^{-1} significantly increased lipid peroxidation levels and up to 31% inhibition of acetylcholinesterase activity. Additionally, the researchers found that in the mixture, there was a decrease in the nominal concentration of chromium (VI) in the aqueous medium, in the presence of microplastics, thereby suggesting that chromium (VI) sorbed to the microplastics when the two substances were mixed (Lusher et al., 2015).

Conversely, in another study which involved gold, rather than chromium IV, it was observed that when the juvenile common goby (*P. microps*) was exposed to ≈5 nm gold nanoparticles for 96 h, there was a reduction in predatory performance of approximately 39%. At the same time, an increase in water temperature from 20°C to 25°C resulted in a 2.3-fold increase from 0.129 to 0.129 g g^{-1} wet weight in the weight of gold nanoparticles incorporated within their body. However, the presence of 1–5 μm polyethylene mini-microplastics during the exposure did not affect the toxicity of gold nanoparticles to the organism (Cózar et al., 2015).

A study of wild gudgeon freshwater fish (*Gobio gobio*) collected from French rivers found microplastics in 12% of the fish. In another study which involved investigating albacore (*Thunnus alalunga*), bluefin tuna (*Thunnus thynnus*), and swordfish (*Xiphias gladius*) collected from the Mediterranean Sea over 1 year for evidence of plastic consumption, it was found that 18.2% of the fish had ingested plastics, with Bluefin tuna had ingested the greatest amounts (Rummel et al., 2016). Furthermore, 75% of the plastics were in the microplastic size range (<5 mm), as opposed to meso­plastics (5–25 mm) or macroplastics (>25 mm) (Schettler, 2006).

An examination of the gut tract of marine and freshwater fish in the Gulf of Mexico revealed that 10% of marine fish and 8% of freshwater fish had ingested microplastics, as well as other sized plastics. Although the plastics found had a maximum size of 14.3 mm, the most common sizes were 1–2 mm. The plastics found were polystyrene, poly(methyl methacrylate), nylon, polyester, and polypropylene. Interestingly, the researchers did not report finding any polyethylene microplastics. Considering that fish were examined and that polyethylene is typically one of the most common types of plastic reported in many other studies, this is quite remarkable. Consequently, other factors, such as color and buoyancy, may be at play. A study of the common goby (*P. microps*) found that as visual predators, the juvenile fish were confused by polyethylene mini-microplastics in the size range 420–500 μm and misidentified them as their natural prey (*Artemia nauplii*), thereby reducing their predatory abilities drastically. Interestingly, the researchers speculate that mini-microplastics moving in the aquatic environment due to turbulence would likely resemble the movements of natural prey. Thus, mini-microplastics would be more susceptible to misidentification and consumption. Certainly, it has been reported that many plastic items collected from marine environments show signs of bite marks by fish. It has been estimated that 1.3 tons of plastic were attacked in a 15 km area near Hawaii (Lithner et al., 2011).

5.3.4 Impact of Microplastics and Nanoplastics on Birds

Birds have more than 10,000 living species among the tetrapod classes (Ducatez and Lefebvre 2014). They are endotherms organisms that are widely distributed in various habitats from the equator to polar areas and from oceans and freshwater to high plateaus. They exhibit flight-related morphological and physiological traits that enable them to occupy different habitats and become important members of many ecosystems (Orme et al., 2006) and are believed to be highly sensitive and vulnerable to external conditions. Compared with non-flying animals, birds have better antioxidant capacity (Costantini, 2008), a higher metabolic rate (McNab, 2009), prolonged lifespan (Munshi-South and Wilkinson, 2010), and short but efficient digestive tract (Caviedes-Vidal et al., 2007).

Several studies have reported that the deposited and aggregated plastics can cause bleeding, blockage of the digestive tract, ulcers, or perforations of the gut, which can produce a deceptive feeling of satiation (Derraik, 2002; Pierce et al., 2004), lead to starvation (Derraik, 2002; Pierce et al., 2004), or cause direct mortality (Derraik, 2002; Roman et al., 2019).

Microplastic fibers, beads, and microplastics have been found embedded in the intestinal wall of Red-shouldered Hawk and Osprey, indicating that these materials can remain in the intestines longer than other indigestible items (Carlin et al., 2020).

A decreased growth rate induced by plastic ingestion was observed in the chicks of Flesh-footed Shearwater (*Puffinus carneipes*) and Japanese Quail (*Coturnix japonica*) (Roman et al., 2019), which likely resulted from reduced stomach capacity rather than toxicological effects.

Some studies have found that ingestion of MPs has reproductive toxicity to birds (Roman et al., 2019; Fossi et al., 2018). For example, chicks of Japanese Quail with observed plastic ingestion exhibited a minor delay in sexual maturity and a higher incidence of epididymal intra-epithelial cysts in males. However, there were no effects on reproductive success (Roman et al., 2019). Similarly, the ingestion of MPs can also reduce the reproductive output of Flesh-footed Shearwater (Fossi et al., 2018).

Furthermore, ingestion of MPs by birds can activate inflammatory responses and lead to reducing food intake, delayed ovulation, and increased mortality (Wright et al., 2013; Carbery et al., 2018; Fossi et al., 2018). In this context, it is important to determine the potential MPs concentration that is detrimental or sublethal to body condition, growth, development, reproduction, and other physiological functions in birds (Puskic et al., 2019).

Studies also indicated that ingestion of toxic substances adsorbed on MPs can induce malnutrition, endocrine disruption, and issues in the reproductive biology of Japanese Quail (Roman et al., 2019) and several species of seabirds, including Kelp Gull (*Larus dominicanus*) (Barbieri et al., 2010), Short-tailed Shearwater (Tanaka et al., 2013), White-chinned Petrel (*Procellaria aequinoctialis*), Slender-billed Prion (*Pachyptila belcheri*), Great Shearwater, Black-browed Albatross (*Thalassarche melanophrys*), and Southern Giant Petrel (*Macronectes giganteus*) (Susanti et al., 2020). Chronic exposure to EDCs can have several negative effects on the developmental and reproductive biology of Japanese Quail, Tree Swallow (*Tachycineta bicolor*) (McCarty and Secord, 2000), American Kestrel (*Falco sparverius*) (Fisher et al., 2001), Great Blue Heron (*Ardea herodias*) and White Ibis (*Eudocimus albus*) (Jayasena et al., 2011), and it also can impair immune and thyroid functions in Japanese Quails. Furthermore, EDCs cause poor reproductive output because of embryonic death, chick deformities, eggshell thinning, and even death in Japanese Quails (Ottinger et al., 2005). Previous studies have shown that traditional pollutants, such as heavy metals and POPs, are detrimental to birds' health. For example, heavy metals have adverse effects on the testicular function, and sperm quality of Eurasian Tree Sparrows (*Passer montanus*) (Yang et al., 2019) and White Ibises (Frederick and Jayasena, 2011), and POPs exert numerous negative effects on endocrine, immune and neural system in White-tailed Eagle (*Haliaeetus albicilla*) (Sletten et al., 2016) and reproduction, and development, and growth in other bird species (Hao et al., 2021). Figures 5.6 and 5.7 depict the impact of environmental plastics on bird and uptake pathways.

5.3.5 IMPACT OF MICROPLASTICS AND NANOPLASTICS ON MAMMALS

5.3.5.1 Impact on Seals and Sea Otters

Higher organisms can be exposed to microplastics and contaminants purely due to their diet. It has been suggested that the exposure of seals to POPs is primarily via

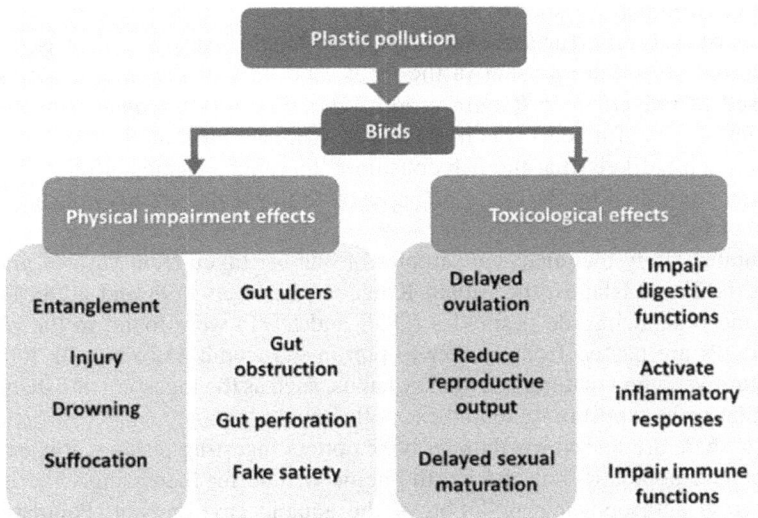

FIGURE 5.6 Impact of environmental plastics on the bird. (Retrieved from Wang et al., 2021 under a Creative Commons License copyright.)

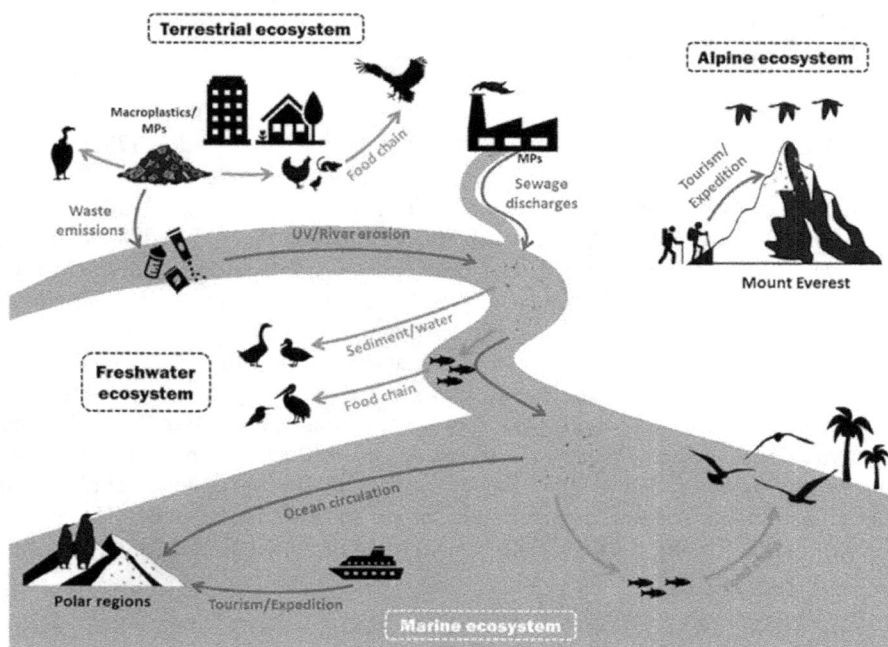

FIGURE 5.7 Cycling process of various plastics in different ecosystems and uptake pathways. (Retrieved from Wang et al., 2021 under a Creative Commons License copyright.)

ingestion of contaminated food sources, such as fish, instead of contact with the aquatic environment (Fisner et al., 2013). For example, in a study that examined the feces of Antarctic fur seals on Macquarie Island, Australia, it was revealed that microplastics were present in the feces, and 93% of the microplastics were composed of polyethylene (Gouin et al., 2011; EU, 2008). Considering that the main diet of these fur seals was mesopelagic fish, such as lantern fish (*Electrona subaspera*), it is likely that the fish consumed the microplastics, after which they were then consumed by the seals, as opposed to the seals directly consuming the microplastics.

In another study that analyzed samples of blubber taken from various gray seal pups in the Farne Islands, the United Kingdom, between 1998 and 2000, PBDEs, PCBs, and organochloride pesticides (DDE and DDT) were found in the blubber. These POPs are passed from mother to pup in the womb and via milk following birth. However, upon maturation, dietary habits, such as the ingestion of fish, become the predominant exposure route to these pollutants.

While there are no reports thus far of sea otters ingesting plastics, it is expected that sea otter populations in the North Pacific will suffer increasingly detrimental effects from anthropogenic pollution of the aquatic environment. Populations in California in the United States have exhibited the highest effects from pollutants, suffering infectious diseases and increased morbidity and thus have been deemed as threatened by the United States Endangered Species Act (Kampire et al., 2015; Ballachey and Bodkin, 2015).

5.3.5.2 Impact on Whales

There have been several documented incidences of many species of whale ingesting plastic litter, such as the killer whale, the beaked whale, and the sperm whale (Shashoua, 2012). Furthermore, a dead sperm whale recovered from the Mediterranean Sea was ascertained to have died as a result of a ruptured stomach, which contained 7.6 kg of plastic material (De Wael et al., 2008). However, microplastics were directly identified for the first time in a cetacean species during a study of True's beaked whale carcasses discovered on Ireland's North and West coasts. In the carcass examined, the average size of the microplastics was reported to be 2.16 mm, while the greatest accumulation of microplastics (38%) was in the whale's main stomach. Furthermore, the main plastic recovered overall was rayon, which comprised 53% of the recovered microplastics. This was significantly higher than other types of plastic and the next most abundant, polyester, at 16%. Indeed, the common plastics polyethylene and polypropylene were reported to be only 4% and 6%, respectively. In another whale, a 7 cm polyethylene shotgun cartridge was found in the main accessory stomach of the whale. However, the researchers stress that ingestion of large plastic debris by these whales is quite common. However, as the largest filter feeder in the ocean, baleen whales (such as the blue whale) were expected to be exposed to large quantities of microplastics daily via their filtering activities (Megharaj et al., 2011). In a recent study, microplastics were identified in a baleen whale (*Megaptera novaeangliae*) carcass on a Netherlands sandbank. The plastic materials identified were mainly fragments, sheets, and threads ranging from 1 to 17 mm. Furthermore, it was reported that the abundance and types of plastic found were polyethylene (55.01%), polyamide (37.64%), polypropylene (5.61%), polyvinyl chloride (0.97%), and polyethylene terephthalate (0.77%) (Apul and Karanfil, 2015).

In a study that analyzed the blubber of stranded baleen whales along the Italian coast, a significantly high concentration (57.97 ng g^{-1}) of mono-(2-ethylhexl) phthalate, a metabolite of di-(2-ethylhexyl) phthalate, was detected in 80% of the blubber. The researchers concluded that whales are chronically exposed to phthalates and possibly other POPs from the regular consumption of plastic litter and have stressed that this population of whales is now in sharp decline. Furthermore, studies of beluga whale carcasses have revealed high levels of POPs upon toxicological examination, and reproductive orders exposed animals (Eerkes-Medrano et al., 2015).

Furthermore, examinations of five species of toothed whales found high concentrations of PBDEs and PCBs in their blubber, with main factors affecting levels of pollutants in these animals considered to be the local environment, POPs containing food sources, and trophic level. Certainly, the levels of POPs within the tissues of organisms tend to increase in concentration at consecutively higher levels in the food chain, thereby demonstrating biomagnification (Fischer et al., 2015).

While POPs which exhibit considerable hydrophobicity, such as PCBs, build up in the tissues of organisms to concentrations that exceed that of the surrounding water (bioconcentrate). Models suggest that compounds such as hexachlorocyclohexane, which exhibit a low octanol–water partition coefficient (K_{ow}) tend not to bioconcentrate in aquatic species. However, biomagnification of β-hexachlorocyclohexane in the fatty tissues of beluga whales was higher than the biomagnification of the PCB congener 2,2′, 3,4,4′, 5,5′-heptachlorobiphenyl (PCB-180), which was attributed to an increased digestive capability of air-breathers to assimilate their food sources, coupled with decreased urinary elimination of POPs with a Log K_{ow} greater than 2. For water-respiring creatures, POPs require a Log K_{ow} greater than five before bioconcentration occurs. Thus, the bioconcentration of POPs may be substantially greater in air-breathing animals than in water-respiring animals (Klein et al., 2005).

5.3.5.3 Impact of Microplastics and Nanoplastics on Human Health

The ubiquitous microplastics can be intake via the exposure pathways (e.g., ingestion, inhalation) and potentially pose a threat to human health (Zhang et al., 2020, Wright and Kelly, 2017). Figure 5.8 indicates the effect of microplastics and nanoplastics on human health.

Dermal contact: Dermal contact with microplastics is considered a less important route of exposure, although it has been speculated that nanoplastics (<100 nm) could transverse the dermal barrier (Revel et al., 2018). Nonetheless, the possibility that nanoplastics can cross the dermal barrier and cause toxicity should not be abandoned without proof. In medicine, plastics are known to induce low inflammatory reactions and a foreign body reaction with fibrous encapsulation. For instance, surgical sutures using braided polyester and monofilament polypropylene resulted in lower inflammatory reactions than silk and fibrous encapsulation after 21 days (Salthouse and Matlaga, 1975). Even though microplastics and nanoplastics could also induce inflammation and foreign body reactions, differences in surface properties could also lead to distinct outcomes. Human epithelial cells suffer oxidative stress from exposure to microplastics and nanoplastics (Schirinzi et al., 2017). Thus, the potential adverse effects of nanoplastics and widespread dermal exposure to plastic particles (i.e., dust, synthetic fibers, and microbeads in cosmetics) need further investigation.

Inhalation, deposition, and translocation: Limited literature exists on potential adverse effects of MPs on human health. The fate of inhaled MPs and their uptake in lung tissue is unknown. For example, there is only one report of polymeric fibers in human lung tissue published more than 20 years ago. In that study, polymeric and cellulosic fibers were found in 97% of malignant lungs ($n = 32/33$) and 83% of non-neoplastic lungs ($n = 67/81$). The analyzed fibers had lengths up to 135 μm and showed little deterioration, which may indicate their bio-resistance and bio-persistence in the lungs. This single study shows that polymeric fibers can penetrate deeper parts of the lungs and highlights the need to confirm these findings and more in-depth analyses. Importantly, studies on, for example, asbestos fibers have shown that despite their length, such fibers can be deposited in the alveolar region of the lung. This result is based on the finding that asbestos fibers with lengths ranging from 50 to 200 μm were found in the alveolar cavity despite their long size. Although the physical characteristics of polymeric and asbestos fibers differ, both are known to resist biodegradation (Wright and Kelly, 2017; Pauly et al., 1998).

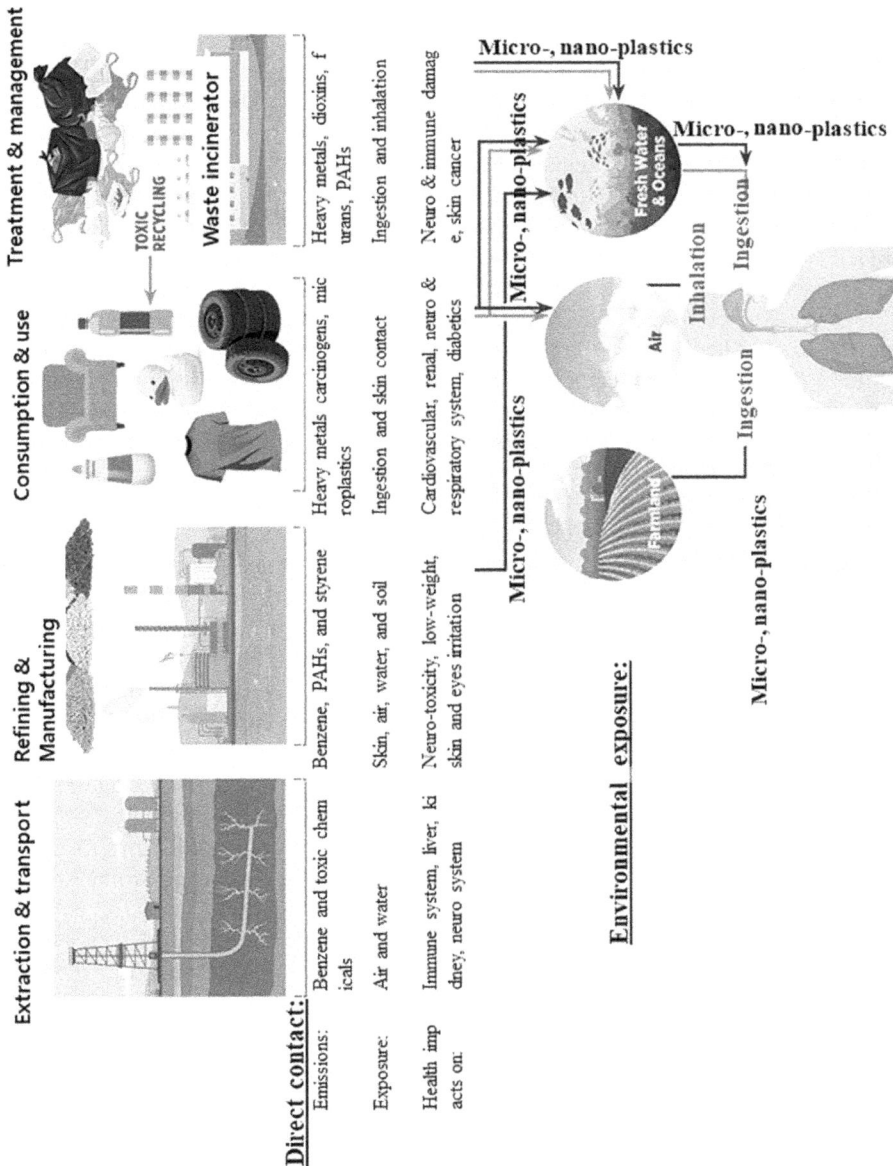

FIGURE 5.8 Impact of microplastics and nanoplastics on human health.

Particle deposition in the lung is a function of its aerodynamic diameter that is particle size expressed in terms of settling speed. Fiber diameter plays a major role in its breathability, while the length is a key determinant of its persistence and toxicity. Pleural mesotheliomas are usually associated with fibers over 8 μm in length and less than 0.25 μm in diameter. Furthermore, it is known that the efficiency of fiber deposition increases with a decrease in diameter (Donaldson et al., 1993).

Fibers can be deposited in terminal bronchioles, alveolar ducts, and alveoli, resulting in chronic inflammation, granulomas, or fibrosis. The severity of tissue damage is usually a function of an inhaled dose over time (Warheit et al., 2001). Greim et al. (2001) suggested that the interaction between cells and particles/fibers can cause inflammation, which induces cell proliferation and secondary genotoxicity due to the continuous formation of reactive oxygen species (ROS). Overproduction of ROS results in oxidative stress, causing chronic inflammation and contributing to the pathogenesis of lung diseases.

Fiber persistence in the lung is related to its aerodynamic properties (Tian and Ahmadi, 2013). The fiber length to diameter ratio determines their uptake by alveolar macrophages and affects mucociliary clearance rates. Usually, long, thin fibers are incompletely phagocytosed and are more biologically active than short fibers (Donaldson et al., 1993). These persistent particles can translocate into the epithelial layers (Donaldson et al., 2011) and induce acute or chronic inflammatory processes.

Occupational exposure to PVC dust was associated with exertional dyspnea and decreased pulmonary function in factory workers (Soutar et al., 1980). A study conducted by Atis et al. (2005) evaluated the respiratory effects of occupational polypropylene flock exposure. The risk of respiratory symptoms increased 3.6-fold in polypropylene flocking workers compared to controls. Lung biopsies from workers exposed to different airborne synthetic fibers (acrylic, polyester (terylene) nylon) revealed different degrees of inflammation, granulomas, and interstitial fibrosis (Pimentel et al., 1975).

These observations from occupational medicine, combined with the recent detection of MPs in airborne samples, point to a possible risk for human exposure via inhalation. These studies suggest that there may be a link between MP exposure and the development of interstitial lung diseases, but further research needs to be done in this direction.

Ingestion; Ingestion is considered the major route of human exposure to microplastics and nanoplastics. Based on foodstuff consumption, the estimated intake of microplastics is 39,000–52,000 particles person^{-1}year^{-1} (Galloway, 2015; Cox et al., 2019). Particles may reach the gastrointestinal system through contaminated foodstuff or the mucociliary clearance after inhalation, possibly leading to inflammatory response, increased permeability, and gut microbe composition and metabolism (Salim et al., 2014). Microplastics have been reported in food items, such as mussels, commercial fish, as well as table salt, sugar, and bottled water, so their ingestion can also occur from these food items (Li et al., 2016; Neves et al., 2015; Karami et al., 2017; Liebezeit and Liebezeit, 2013; Oßmann et al., 2018).

The study of Stock and colleagues revealed no striking alteration of the physico-chemical characteristics of five types of MPs (PE, PP, PVC, PET, and PS) by artificial digestive juices, mimicking the saliva, gastric, and intestinal phases of human digestion. All digestion steps were successive in the same vessel. MPs were digested for 5 min at 37°C in the presence of synthetic saliva containing α-amylase, mucin, and various electrolytes, under agitation at pH 6.7. Gastric juice, composed of pepsin, mucin, electrolytes, and hydrochloride acid, was then added, and the pH was adjusted at 2 for the gastric phase for 2 h. Finally, the intestinal digestion began with the addition of artificial intestinal juice containing trypsin, pancreatin, bile extract, and electrolytes, with pH value set at 7.5 for 2 h. The same authors emphasized the importance of considering corona formation onto MPs surface due to the adsorption of organic compounds such as proteins, mucins, and lipids during the digestion process. Owing to its biological impact the characteristics of the biomolecular corona formed around the MPs during their transit in the digestive tract need to be more deeply studied to determine the behavior of the MPs in the gut, and particularly their uptake rate (Monopoli et al., 2012).

Besides MPs themselves, other contaminants adsorbed or entangled with MPs can also harm human health. Liao and Yang (2020) reported high desorption of Cr, as a contaminant with MPs, under acidic conditions in a human static *in vitro* digestion model, including the mouth, gastric, small intestinal, and large intestinal digestive phases. The digestion was performed at 37°C on glass tubes for each compartment independently with adapted juices composition, digestion time, and pH. It was observed that Cr adsorbed on various MP types (PE, PP, PVC, PS, and PLA, 150 μm) was more bioaccessible in the gastric environment than in the small or large intestinal phases, due to acidic conditions favoring desorption of anionic Cr species (e.g., CrO_4^{2-} and $HCrO_4^-$) from the MP surface, while no release was observed in the oral phase. Moreover, PLA exhibited the highest oral bioaccessibility of Cr(VI) and Cr(III) in comparison to other polymer types, probably due to its degradation enhanced by the action of enzymes present in simulated digestive juices.

Similarly, microorganisms, including bacteria, are able to aggregate and adhere to biological or non-biological surfaces in complex communities commonly referred to as biofilms (López et al., 2010). But literature is not available so far to unravel the consequences of oral exposure to pathogen- and/or antibiotic resistance gene-contaminated MPs during *in vitro* human digestion or *in vivo* in mammals except reported concentrations of antibiotics absorbed onto MPs. Specifically, levofloxacin (fluoroquinolone) can be adsorbed in concentrations up to 1.2 mg g^{-1} (Yu et al., 2019) onto PVC, whereas ciprofloxacin (fluoroquinolone), amoxicillin (penicillin), and tetracycline can be adsorbed to polyamide (PA) in concentrations ranging from 1 to 3 mg g^{-1} (Li et al., 2018). Sulfamethoxazole was found to be strongly adsorbed onto PA and PE MPs in concentrations of 0.4 and 0.1 mg g^{-1}, respectively (Guo et al., 2019, Razanajatovo et al., 2018). These results point out that some of the most common MP particles are vectors for potential pathogenic microbes and favor hitch-hiking of microbes that are resistant to antimicrobials. Those may facilitate the transfer of antimicrobial resistance genes to the intestinal microbiome.

Microorganisms may also colonize the surface of microplastics, including *Vibrio* spp. (Kirstein et al., 2016). In this case, microplastics could act as vectors, delivering microorganisms to the tissues, protecting them from the immune system, and creating tissue damage that may favor infection. Furthermore, microplastics altered and increased the diversity of the gut microbiome in soil organisms (*Folsomia candida*). The same effect could happen in humans after ingesting a significant amount of microplastics. Alterations to the gut microbiome could lead to adverse effects, such as the proliferation of harmful species, **an** increase in intestinal permeability, and endotoxemia (West-Eberhard, 2019). Nonetheless, negative effects from the release of chemicals or microorganisms adsorbed to microplastics will be highly dependent on the types associated with ingested particles, the clearance time and translocation of vector microplastics, the release rate, and the extent of the contaminant, and its translocation and noxious effects in human tissues.

ACKNOWLEDGMENTS

This work was supported by a grant from the National Research Foundation of Korea (NRF) grant funded by the Korea government (MSIT) (No. NRF-2020R1A2C1013851).

REFERENCES

Apul, O. G., Karanfil, T. (2015) Adsorption of synthetic organic contaminants by carbon nanotubes: a critical review. *Water Research*, 68, 34–55.

Atis, S., Tutluoglu, B., Levent, E., Ozturk, C., Tunaci, A., Sahin, K., Nemery, B. (2005) The respiratory effects of occupational polypropylene flock exposure. *European Respiratory Journal*, 25(1), 110–117.

Au, S. Y., Bruce, T. F., Bridges, W. C., Klaine, S. J. (2015) Responses of Hyalella azteca to acute and chronic microplastic exposures. *Environmental Toxicology and Chemistry*, 34(11), 2564–2572.

Avio, C. G., Gorbi, S., Milan, M., Benedetti, M., Fattorini, D., d'Errico, G., Regoli, F. (2015) Pollutants bioavailability and toxicological risk from microplastics to marine mussels. *Environmental Pollution*, 198, 211–222.

Ballachey, B. E., Bodkin, J. L. (2015) Challenges to sea otter recovery and conservation. *Sea Otter Conservation*. DOI:10.1016/B978-0-12-801402-8.00004-4.

Barbieri, E., de Andrade Passos, E., Filippini, A., dos Santos, I. S., Garcia, C. A. B. (2010) Assessment of trace metal concentration in feathers of seabird (Larus dominicanus) sampled in the Florianópolis, SC, Brazilian coast. *Environmental Monitoring and Assessment*, 169(1), 631–638.

Bergami, E., Bocci, E., Vannuccini, M. L., Monopoli, M., Salvati, A., Dawson, K. A., Corsi, I. (2016) Nano-sized polystyrene affects feeding, behavior and physiology of brine shrimp Artemia franciscana larvae. *Ecotoxicology and Environmental Safety*, 123, 18–25.

Betts, K. (2008) Why small plastic particles may pose a big problem in the oceans. *Environmental Science and Technology*, 42(24), 8995.

Brillant, M., MacDonald, B. (2002) Postingestive selection in the sea scallop (*Placopecten magellanicus*) on the basis of chemical properties of particles. *Marine Biology*, 141(3), 457–465.

Browne, M. A., Galloway, T. Thompson, R. (2007) Microplastic—an emerging contaminant of potential concern? *Integrated Environmental Assessment and Management: An International Journal*, 3(4), 559–561.

Browne, M. A., Niven, S. J., Galloway, T. S., Rowland, S. J., Thompson, R. C. (2013) Microplastic moves pollutants and additives to worms, reducing functions linked to health and biodiversity. *Current Biology*, 23(23), 2388–2392.

Carbery, M., O'Connor, W.,Palanisami, T. (2018) Trophic transfer of microplastics and mixed contaminants in the marine food web and implications for human health. *Environment International*, 115, 400–409.

Carlin, J., Craig, C., Little, S., Donnelly, M., Fox, D., Zhai, L., Walters, L. (2020) Microplastic accumulation in the gastrointestinal tracts in birds of prey in central Florida, USA. *Environmental Pollution*, 264, 114633.

Caviedes-Vidal, E., McWhorter, T. J., Lavin, S. R., Chediack, J. G., Tracy, C. R., Karasov, W. H. (2007) The digestive adaptation of flying vertebrates: high intestinal paracellular absorption compensates for smaller guts. *Proceedings of the National Academy of Sciences*, 104(48), 19132–19137.

Claessens, M., Van Cauwenberghe, L., Vandegehuchte, M. B., Janssen, C. R. (2013) New techniques for the detection of microplastics in sediments and field collected organisms. *Marine Pollution Bulletin*, 70(1–2), 227–233.

Cole, M., Lindeque, P., Halsband, C., Galloway, T. S. (2011) Microplastics as contaminants in the marine environment: a review. *Marine Pollution Bulletin*, 62(12), 2588–2597.

Cole, M., Galloway, T. S (2015) Ingestion of nanoplastics and microplastics by pacific oyster larvae. *Environmental Science & Technology*, 49(24), 14625–14632.

Costantini, D. (2008) Oxidative stress in ecology and evolution: lessons from avian studies. *Ecology Letters*, 11(11), 1238–1251.

Cox, K. D., Covernton, G. A., Davies, H. L., Dower, J. F., Juanes, F., Dudas, S. E. (2019) Human consumption of microplastics. *Environmental Science & Technology*, 53(12), 7068–7074.

Cózar, A., Sanz-Martín, M., Martí, E., González-Gordillo, J. I., Ubeda, B., Gálvez, J. Á., Irgoien, X., Duarte, C. M. (2015) Plastic accumulation in the Mediterranean sea. *PLoS One* 10(4), 1–12.

da Costa Araújo, A. P., Malafaia, G. (2020) Can short exposure to polyethylene microplastics change tadpoles' behavior? A study conducted with neotropical tadpole species belong to order anura (*Physalaemus cuvieri*). *Journal of Hazardous Materials*, 391, 122214.

Daly, G. L., Wania, F. 2005. Organic contaminants in mountains. *Environmental Science & Technology* 39(2), 385–398.

de Sá, L. C., Luís, L. G., Guilhermino, L. (2015) Effects of microplastics on juveniles of the common goby (*Pomatoschistus microps*): confusion with prey, reduction of the predatory performance and efficiency, and possible influence of developmental conditions. *Environmental Pollution*, 196, 359–362.

Carlos de Sá, L., Oliveira, M., Ribeiro, F., Rocha, T. L., Futter, M. N. (2018) Studies of the effects of microplastics on aquatic organisms: What do we know and where should we focus our efforts in the future?, *Science of the Total Environment*, 645, 1029–1039.

de Souza Machado, A. A., Lau, C. W., Kloas, W., Bergmann, J., Bachelier, J. B., Faltin, E., Becker, R., Görlich, A. S., Rillig, M.C.. (2019) Microplastics can change soil properties and affect plant performance. *Environmental Science and Technology*, 53, 6044–6052.

De Wael, K., Gason, F. G., Baes, C. A. (2008) Selection of an adhesive tape suitable for forensic fiber sampling. *Journal of Forensic Sciences*, 53(1), 168–171.

Della Torre, C., Bergami, E., Salvati, A., Faleri, C., Cirino, P., Dawson, K. A., Corsi, I. (2014) Accumulation and embryotoxicity of polystyrene nanoparticles at early stage of development of sea urchin embryos Paracentrotus lividus. *Environmental Science & Technology*, 48(20), 12302–12311.

Derraik, J. G. (2002) The pollution of the marine environment by plastic debris: a review. *Marine Pollution Bulletin*, 44(9), 842–852.

Desforges, J-P. W., Galbraith, M., Dangerfield, N., Ross, P. S. (2014) Widespread distribution of microplastics in subsurface seawater in the NE Pacific Ocean. *Marine Pollution Bulletin*, 79(1–2), 94–99.

Desforges, J. P. W., Galbraith, M., Ross, P. S. (2015) Ingestion of microplastics by zooplankton in the Northeast Pacific Ocean. *Archives of Environmental Contamination and Toxicology*, 69(3), 320–330.

Donaldson, K., Brown, R.C., Brown, G. M (1993) Respirable industrial fibres: mechanisms of pathogenicity. *Thorax*, 48, 390–395.

Donaldson, K., Murphy, F., Schinwald, A., Duffin, R., Poland, C. A. (2011) Identifying the pulmonary hazard of high aspect ratio nanoparticles to enable their safety-by-design. *Nanomedicine*, 6(1), 143–156.

Douabul, A. A., Heba, H. M., Fareed, K. H. (1997) Polynuclear aromatic hydrocarbons (PAHs) in fish from the Red Sea Coast of Yemen. Asia-Pacific Conference on Science and Management of Coastal Environment. *Developments in Hydrobiology*, 123, 251–262.

Ducatez, S., Lefebvre, L. (2014) Patterns of research effort in birds. *PLoS One*, 9(2), e89955.

Eerkes-Medrano, D., Thompson, R. C., Aldridge, D. C. (2015) Microplastics in freshwater systems: a review of the emerging threats, identification of knowledge gaps and prioritisation of research needs. *Water Research*, 75, 63–82.

Erdman, Jr, J. W., MacDonald, I. A., Zeisel, S. H. (2012) *Present Knowledge in Nutrition*. John Wiley & Sons, Hoboken, New Jersey, US.

EU. (2008) Directive 2008/56/EC of the European Parliament and of the Council of 17 June 2008 establishing a framework for community action in the field of marine environmental policy (Marine Strategy Framework Directive). EU, Brussels.

Fischer, V., Elsner, N. O., Brenke, N., Schwabe, E., Brandt, A. (2015) Plastic pollution of the Kuril–Kamchatka Trench area (NW pacific). *Deep Sea Research Part II: Topical Studies in Oceanography*, 111, 399–405.

Fisher, S. A., Bortolotti, G. R., Fernie, K. J., Smits, J. E., Marchant, T. A., Drouillard, K. G., & Bird, D. M. (2001) Courtship behavior of captive American kestrels (Falco sparverius) exposed to polychlorinated biphenyls. *Archives of Environmental Contamination and Toxicology*, 41(2), 215–220.

Fisner, M., Taniguchi, S., Moreira, F., Bícego, M. C., Turra, A. (2013) Concentration and composition of polycyclic aromatic hydrocarbons (PAHs) in plastic pellets: implications for small-scale diagnostic and environmental monitoring. *Marine Pollution Bulletin*, 76(1–2), 349–354.

Fossi, M. C., Panti, C., Baini, M., Lavers, J. L. (2018) A review of plastic-associated pressures: cetaceans of the Mediterranean Sea and eastern Australian shearwaters as case studies. *Frontiers in Marine Science*, 5, 173.

Fossi, M. C., Panti, C., Guerranti, C., Coppola, D., Giannetti, M., Marsili, L., Minutoli, R. (2012) Are baleen whales exposed to the threat of microplastics? A case study of the Mediterranean fin whale (*Balaenoptera physalus*). *Marine Pollution Bulletin*, 64(11), 2374–2379.

Frederick, P., Jayasena, N. (2011) Altered pairing behaviour and reproductive success in white ibises exposed to environmentally relevant concentrations of methylmercury. *Proceedings of the Royal Society B: Biological Sciences*, 278, 1851–1857.

Galloway, T. S. (2015) Micro-and nano-plastics and human health. In *Marine Anthropogenic Litter*. Springer, Cham. pp. 343–366.

Gambardella, C., Morgana, S., Ferrando, S., Bramini, M., Piazza, V., Costa, E., Faimali, M. (2017) Effects of polystyrene microbeads in marine planktonic crustaceans. *Ecotoxicology and Environmental Safety*, 145, 250–257.

Gouin, T., Roche, N., Lohmann, R., Hodges, G. (2011) A Thermodynamic Approach for Assessing the Environmental Exposure of Chemicals Absorbed to Microplastic. *Environmental Science & Technology*, 45(4), 1466–1472.

Graham, E. R., Thompson, J. T. (2009) Deposit-and suspension-feeding sea cucumbers (Echinodermata) ingest plastic fragments. *Journal of Experimental Marine Biology and Ecology*, 368(1), 22–29.

Greim, H., Borm, P., Schins, R., Donaldson, K., Driscoll, K., Hartwig, A., Speit, G. (2001) Toxicity of fibers and particles? Report of the workshop held in Munich, Germany, 26? 27 October 2000. *Inhalation Toxicology*, 13(9), 737–754.

Guo, X., Liu, Y., Wang, J. (2019) Sorption of sulfamethazine onto different types of microplastics: a combined experimental and molecular dynamics simulation study. *Marine Pollution Bulletin*, 145, 547–554.

Hao, Y., Zheng, S., Wang, P., Sun, H., Matsiko, J., Li, W., Jiang, G. (2021) Ecotoxicology of persistent organic pollutants in birds. *Environmental Science: Processes & Impacts*, 23(3), 400–416.

Hart, M. W. (1991) Particle captures and the method of suspension feeding by echinoderm larvae. *The Biological Bulletin*, 180(1), 12–27.

Herzke, D., Anker-Nilssen, T., Nøst, T. H., Götsch, A., Christensen-Dalsgaard, S., Langset, M., Koelmans, A. A. (2016) Negligible impact of ingested microplastics on tissue concentrations of persistent organic pollutants in northern fulmars off coastal Norway. *Environmental Science & Technology*, 50(4), 1924–1933.

Hurley, R., Woodward, J., Rothwell, J. J. (2018) Microplastic contamination of river beds significantly reduced by catchment-wide flooding. *Nature Geoscience*, 11(4), 251–257.

Ivar do Sul, J. A., Costa, M. F. (2014) The present and future of microplastic pollution in the marine environment. *Environmental Pollution*, 185, 352–364.

Jayasena, N., Frederick, P. C., Larkin, I. L. (2011) Endocrine disruption in white ibises (Eudocimus albus) caused by exposure to environmentally relevant levels of methylmercury. *Aquatic Toxicology*, 105(3–4), 321–327.

Jeong, C. B., Won, E. J., Kang, H. M., Lee, M. C., Hwang, D. S., Hwang, U. K., Lee, J. S. (2016) Microplastic size-dependent toxicity, oxidative stress induction, and p-JNK and p-p38 activation in the monogonont rotifer (*Brachionus koreanus*). *Environmental Science & Technology*, 50(16), 8849–8857.

Kampire, E., Rubidge, G., Adams, J. B. (2015) Distribution of polychlorinated biphenyl residues in sediments and blue mussels (*Mytilus galloprovincialis*) from Port Elizabeth Harbour, South Africa. *Marine Pollution Bulletin*, 91(1), 173–179.

Karami, A., Golieskardi, A., Choo, C. K., Larat, V., Galloway, T. S., Salamatinia, B. (2017) The presence of microplastics in commercial salts from different countries. *Scientific Reports*, 7(1), 1–11.

Kirstein, I. V., Kirmizi, S., Wichels, A., Garin-Fernandez, A., Erler, R., Löder, M., Gerdts, G. (2016) Dangerous hitchhikers? Evidence for potentially pathogenic Vibrio spp. on microplastic particles. *Marine Environmental Research*, 120, 1–8.

Klein, E., Lukeš, V., Cibulková, Z. (2005) On the energetics of phenol antioxidants activity. *Petroleum & Coal*, 47, 33–39.

Lenz, R., Enders, K., Stedmon, C. A., Mackenzie, D. M. A., Nielsen, T. G. (2015) A critical assessment of visual identification of marine microplastic using Raman spectroscopy for analysis improvement. *Marine Pollution Bulletin*, 100(1), 82–91.

Li, J., Huihui Liu, H., Chen, J.P. (2018) Microplastics in freshwater systems: A review on occurrence, environmental effects, and methods for microplastics detection. *Water Research*, 137, 362–374.

Li, J., Qu, X., Su, L., Zhang, W., Yang, D., Kolandhasamy, P., Shi, H. (2016) Microplastics in mussels along the coastal waters of China. *Environmental Pollution*, 214, 177–184.

Liao, Y. L., Yang, J. Y. (2020) Microplastic serves as a potential vector for Cr in an in-vitro human digestive model. *Science of the Total Environment*, 703, 134805.

Liebezeit, G., Liebezeit, E. (2013). Non-pollen particulates in honey and sugar. *Food Additives & Contaminants: Part A*, 30(12), 2136–2140.

Lithner, D., Larsson, A., Dave, G. (2011) Environmental and health hazard ranking and assessment of plastic polymers based on chemical composition. *Science of the Total Environment*, 409(18), 3309–3324.

Löder, M., Gerdts, G. (2015) Methodology used for the detection and identification of microplastics-A critical appraisal. In *Marine Anthropogenic Litter*. Springer. ISBN 978-3-319-16509-7.

López, D., Vlamakis, H., Kolter, R. (2010) Biofilms. *Cold Spring Harbor Perspectives in Biology*, 2, a000398.

Lusher, A. (2015) Microplastics in the marine environment: Distribution, interactions and effects. In *Marine Anthropogenic Litter*, Bergmann, M., Gutow, L., Klages, M., Eds., Springer, New York, NY, US, pp. 245–307.

Monopoli, M.P., Åberg, C., Salvati, A., Dawson, K.A. (2012) Biomolecular coronas provide the biological identity of nanosized materials. *Nature Nanotechnology*, 7, 779–786.

Mariussen, E. (2012) Neurotoxic effects of perfluoroalkylated compounds: mechanisms of action and environmental relevance. *Archives of Toxicology*, 86(9), 1349–1367.

Mathalon, A., Hill, P. (2014) Microplastic fibers in the intertidal ecosystem surrounding Halifax Harbor, Nova Scotia. *Marine Pollution Bulletin*, 81(1), 69–79.

McCarty, J. P., Secord, A. L. (2000) Possible effects of PCB contamination on female plumage color and reproductive success in Hudson River tree swallows. *The Auk*, 117(4), 987–995.

McNab, B. K. (2009) Ecological factors affect the level and scaling of avian BMR. *Comparative Biochemistry and Physiology Part A: Molecular & Integrative Physiology*, 152(1), 22–45.

Megharaj, M., Ramakrishnan, B., Venkateswarlu, K., Sethunathan, N., Naidu, R. (2011) Bioremediation approaches for organic pollutants: a critical perspective. *Environment International*, 37(8), 1362–1375.

Munshi-South, J., Wilkinson, G. S. (2010) Bats and birds: exceptional longevity despite high metabolic rates. *Ageing Research Reviews*, 9(1), 12–19.

Neves, D., Sobral, P., Ferreira, J. L., Pereira, T. (2015) Ingestion of microplastics by commercial fish off the Portuguese coast. *Marine Pollution Bulletin*, 101(1), 119–126.

Norén, F., Naustvoll, L. J. (2010) Survey of microscopic anthropogenic particles in Skagerrak. *Klima- og forurensningsdirektoratet Norge Report TA*, 2779(2011), 1–20.

Orme, C. D. L., Davies, R. G., Olson, V. A., Thomas, G. H., Ding, T. S., Rasmussen, P. C., Gaston, K. J. (2006) Global patterns of geographic range size in birds. *PLoS Biology*, 4(7), e208.

OSPAR. (2010) *Guideline for Monitoring Marine Litter on the Beaches in the OSPAR Maritime Area*. OSPAR Commission, London, UK.

Oßmann, B. E., Sarau, G., Holtmannspötter, H., Pischetsrieder, M., Christiansen, S. H., Dicke, W. (2018) Small-sized microplastics and pigmented particles in bottled mineral water. *Water Research*, 141, 307–316.

Ottinger, M. A., Quinn Jr, M. J., Lavoie, E., Abdelnabi, M. A., Thompson, N., Hazelton, J. L., Jaber, M. (2005) Consequences of endocrine disrupting chemicals on reproductive endocrine function in birds: establishing reliable end points of exposure. *Domestic Animal Endocrinology*, 29(2), 411–419.

Oliveira, M., Ribeiro, A., Hylland, K., & Guilhermino, L. (2013, November) Single and combined effects of microplastics and pyrene on juveniles (0+ group) of the common goby Pomatoschistus microps (Teleostei, Gobiidae). *Ecological Indicators*. Elsevier BV. https://doi.org/10.1016/j.ecolind.2013.06.019.

Pauly, J. L., Stegmeier, S. J., Allaart, H. A., Cheney, R. T., Zhang, P. J., Mayer, A. G., Streck, R. J. (1998) Inhaled cellulosic and plastic fibers found in human lung tissue. *Cancer Epidemiology and Prevention Biomarkers*, 7(5), 419–428.

Pierce, K. E., Harris, R. J., Larned, L. S., Pokras, M. A. (2004) Obstruction and starvation associated with plastic ingestion in a Northern Gannet Morus bassanus and a Greater Shearwater Puffinus gravis. *Marine Ornithology*, 32, 187–189.

Pimentel, J. C., Avila, R., Lourenço, A. G. (1975) Respiratory disease caused by synthetic fibres: a new occupational disease. *Thorax*, 30(2), 204–219.

Plastics Europe. (2015) An analysis of European plastics production, demand and waste data. In *Plastics–The Facts*. Association of Plastic Manufacturers, Belgium. pp. 1–30. https://plasticseurope.org/knowledge-hub/plastics-the-facts-2021.

Prata, J. C., da Costa, J. P., Lopes, I., Duarte, A. C., Rocha-Santos, T. (2020) Environmental exposure to microplastics: an overview on possible human health effects. *Science of the Total Environment*, 702, 134455.

Puskic, P. S., Lavers, J. L., Adams, L. R., Grünenwald, M., Hutton, I., Bond, A. L. (2019) Uncovering the sub-lethal impacts of plastic ingestion by shearwaters using fatty acid analysis. *Conservation Physiology*, 7(1), coz017.

Razanajatovo, R. M., Ding, J., Zhang, S., Jiang, H., Zou, H. (2018) Sorption and desorption of selected pharmaceuticals by polyethylene microplastics. *Marine Pollution Bulletin*, 136, 516–523.

Reisser, J., Shaw, J., Wilcox, C., Hardesty, B. D., Proietti, M., Thums, M., Pattiaratchi, C. (2013) Marine plastic pollution in waters around Australia: characteristics, concentrations, and pathways. *PLoS One*, 8(11), e80466.

Revel, M., Châtel, A., Mouneyrac, C. (2018) Micro (nano) plastics: a threat to human health? *Current Opinion in Environmental Science & Health*, 1, 17–23.

Ribeiro, F., Garcia, A. R., Pereira, B. P., Fonseca, M., Mestre, N. C., Fonseca, T. G., Bebianno, M. J. (2017) Microplastics effects in Scrobicularia plana. *Marine Pollution Bulletin*, 122(1–2), 379–391.

Roman, L., Lowenstine, L., Parsley, L. M., Wilcox, C., Hardesty, B. D., Gilardi, K., Hindell, M. (2019) Is plastic ingestion in birds as toxic as we think? Insights from a plastic feeding experiment. *Science of the Total Environment*, 665, 660–667.

Rummel, C. D., Löder, M. G. J., Fricke, N. F., Lang, T., Griebeler, E-M., Janke, M., Gerdts, G. (2016) Plastic ingestion by pelagic and demersal fish from the North Sea and Baltic Sea. *Marine Pollution Bulletin*, 102(1), 134–141.

Sala, B., Balasch, A., Eljarrat, E., Cardona, L. (2021) First study on the presence of plastic additives in loggerhead sea turtles (*Caretta caretta*) from the Mediterranean Sea. *Environmental Pollution*, 283, 117108.

Salim, S. Y., Kaplan, G. G., Madsen, K. L. (2014) Air pollution effects on the gut microbiota: a link between exposure and inflammatory disease. *Gut Microbes*, 5(2), 215–219.

Salthouse, T. N., Matlaga, B. F. (1975) Significance of cellular enzyme activity at nonabsorbable suture implant sites: silk, polyester, and polypropylene. *Journal of Surgical Research*, 19(2), 127–132.

Schettler, T. (2006) Human exposure to phthalates via consumer products. *International Journal of Andrology*, 29(1), 134–139.

Schirinzi, G. F., Pérez-Pomeda, I., Sanchís, J., Rossini, C., Farré, M., Barceló, D. (2017) Cytotoxic effects of commonly used nanomaterials and microplastics on cerebral and epithelial human cells. *Environmental Research*, 159, 579–587.

Shashoua, Y. (2012) *Conservation of Plastics*. Elsevier Ltd, Burlington, VT. ISBN 978-0-7506-6495-0.

Shipp, L. E., Hamdoun, A. (2012) ATP-binding cassette (ABC) transporter expression and localization in sea urchin development. *Developmental Dynamics*, 241(6), 1111–1124.

Sletten, S., Bourgeon, S., Bårdsen, B. J., Herzke, D., Criscuolo, F., Massemin, S., Bustnes, J. O. (2016) Organohalogenated contaminants in white-tailed eagle (Haliaeetus albicilla) nestlings: an assessment of relationships to immunoglobulin levels, telomeres and oxidative stress. *Science of the Total Environment*, 539, 337–349.

Soutar, C. A., Copland, L. H., Thornley, P. E., Hurley, J. F., Ottery, J., Adams, W. G., Bennett, B. (1980) Epidemiological study of respiratory disease in workers exposed to polyvinylchloride dust. *Thorax*, 35(9), 644–652.

Susanti, N. K. Y., Mardiastuti, A., Wardiatno, Y. (2020) Microplastics and the impact of plastic on wildlife: a literature review. *In IOP Conference Series: Earth and Environmental Science*, 528(1), 012013.

Tanaka, K., Takada, H., Yamashita, R., Mizukawa, K., Fukuwaka, M. A., Watanuki, Y. (2013) Accumulation of plastic-derived chemicals in tissues of seabirds ingesting marine plastics. *Marine Pollution Bulletin*, 69(1–2), 219–222.

Thompson, R. C., Olsen, Y., Mitchell, R. P., Davis, A., Rowland, S. J., John, A. W. & Russell, A. E. (2004) Lost at sea: where is all the plastic? *Science*, 304(5672), 838–838.

Tian, L., Ahmadi, G. (2013) Fiber transport and deposition in human upper tracheobronchial airways. *Journal of Aerosol Science*, 60, 1–20.

Van Cauwenberghe, L., Janssen, C. R. (2014) Microplastics in bivalves cultured for human consumption. *Environmental Pollution*, 193, 65–70.

Venâncio, C., Savuca, A., Oliveira, M., Martins, M. A., Lopes, I. (2021) Polymethylmethacrylate nanoplastics effects on the freshwater cnidarian Hydra viridissima. *Journal of Hazardous Materials*, 402, 123773.

Von Moos, N., Burkhardt-Holm, P., Kohler, A. (2012) Uptake and effects of microplastics on cells and tissue of the blue mussel Mytilus edulis L. after an experimental exposure. *Environmental Science & Technology*, 46(20), 11327–11335.

Wang, L., Nabi, G., Yin, L., Wang, Y., Li, S., Hao, Z., Li, D. (2021) Birds and plastic pollution: recent advances. *Avian Research*, 12(1), 1–9.

Warheit, D. B., Hart, G. A., Hesterberg, T. W., Collins, J. J., Dyer, W. M., Swaen, G. M. H., Kennedy, G. L. (2001) Potential pulmonary effects of man-made organic fiber (MMOF) dusts. *Critical Reviews in Toxicology*, 31(6), 697–736.

Webster, L., Walsham, P., Russell, M., Hussy, I., Neat, F., Dalgarno, E., Packer, G., Scurfield, J. A., Moffat, C. F. (2011) Halogenated persistent organic pollutants in deep water fish from waters to the west of Scotland. *Chemosphere*, 83(6), 839–850.

Wei, C. L., Rowe, G. T., Nunnally, C., Wicksten, M. K. (2012) Anthropogenic "Litter" and macrophyte detritus in the deep Northern Gulf of Mexico. *Marine Pollution Bulletin*, 64(5), 966–973.

Wesch, C., Barthel, A-K., Braun, U., Klein, R., Paulus, M. (2016) No microplastics in benthic eelpout (*Zoarces viviparus*): an urgent need for spectroscopic analyses in microplastic detection. *Environmental Research*, 148, 36–38.

West-Eberhard, M. J. (2019) Nutrition, the visceral immune system, and the evolutionary origins of pathogenic obesity. *Proceedings of the National Academy of Sciences*, 116(-3), 723–731.

Wright, S. L., Kelly, F. J. (2017) Plastic and human health: a micro issue? *Environmental Science & Technology*, 51(12), 6634–6647.

Wright, S. L., Thompson, R. C., Galloway, T. S. (2013) The physical impacts of microplastics on marine organisms: a review. *Environmental Pollution*, 178, 483–492.

Wu, Z., Liu, G., Song, S., Pan, S. (2014) Regeneration and recycling of waste thermosetting plastics based on mechanical thermal coupling fields. *International Journal of Precision Engineering and Manufacturing*, 15(12), 2639–2647.

Xu, S., Zhang, L., Lin, Y., Li, R. (2012) Layered double hydroxides used as flame retardant for engineering plastic acrylonitrile-butadiene-styrene (ABS). *Journal of Physics and Chemistry of Solids*, 73(12), 1514–1517.

Xu, X. Y., Lee, W. T., Chan, A. K. Y., Lo, H. S., Shin, P. K. S., Cheung, S. G. (2017) Microplastic ingestion reduces energy intake in the clam Atactodea striata. *Marine Pollution Bulletin*, 124(2), 798–802.

Yang, Y., Zhang, Y., Ding, J., Ai, S., Guo, R., Bai, X., Yang, W. (2019) Optimal analysis conditions for sperm motility parameters with a CASA system in a passerine bird, Passer montanus. *Avian Research*, 10(1), 1–10.

Yu, F., Yang, C., Zhu, Z., Bai, X., Ma, J. (2019) Adsorption behavior of organic pollutants and metals on micro/nanoplastics in the aquatic environment. *Science of the Total Environment*, 694, 133643.

Zhang, Q., Xu, E. G., Li, J., Chen, Q., Ma, L., Zeng, E. Y., Shi, H. (2020) A review of micro-plastics in table salt, drinking water, and air: direct human exposure. *Environmental Science & Technology*, 54(7), 3740–3751.

6 Collection of Microplastics and Nanoplastics from Various Environments and Associated Challenges

Hyunjung Kim, Sadia Ilyas, Gilsang Hong
Hanyang University

Byoung-cheun Lee and Geunbae Kim
National Institute of Environmental Research

CONTENTS

DOI: 10.1201/9781003200628-6

The abundance of microplastics is commonly presented as a numerical or mass concentration. However, with water samples, this is expressed as the weight or number of microplastics per area such as km^2 for sea surface samples or per volume such as m^3 for water columns. For sediments, the abundance of microplastics is expressed as weight or number of microplastics per sediment area or weight like wet weight or dry weight, as well as volume (mL or L). Consequently, this wide variation in the way in which the abundance of microplastics is quantified means that comparisons between studies can often be very difficult (Wheatley et al., 1993; Vesilind, 2003; Zhang et al., 2015; EC, 2013; Hidalgo-Ruz and Thiel, 2013; Crawford and Quinn, 2017).

While there is no standardization for collecting microplastics, the standardized size and color sorting system for the effective categorizing of microplastics, based on their size and appearance, was introduced by Crawford and Quinn (2017).

There are not yet widely accepted standards for sample collection, laboratory analyses, quality assurance/quality control or reporting of microplastics in the environmental samples.

Sampling strategies can be distinguished into selective, bulk, and volume-reduced samplings. Every sampling strategy has its own merits and demerits. Microplastics and nanoplastics in the aquatic environment are sampled using different devices. Sampling devices can be divided into three categories: non-discrete sampling devices include nets and pumping systems; discrete sampling devices include Niskin bottles, rosette, integrating water sampler, bucket, and steel sampler; and sampling devices of the surface microlayer include sieves and rotating drum sampler. The sediment samples are mostly collected by manual picking or using a different corer, while atmospheric samplings are collected by active or passive pumping.

By and large, the sampling method and devices used for sampling depend upon the environment of the sample (i.e., rainfall, wind direction, time of sampling), type of the sample (aquatic, atmospheric, sediments, etc.), and the size limitation of microplastics.

6.1   SAMPLING METHODS FOR MICROPLASTICS AND NANOPLASTICS

The three sampling methods involving selective sampling, volume-reduced sampling, and bulk sampling is mainly practiced, and each has its own merits and demerits (Hidalgo-Ruz et al., 2012).

6.1.1   SELECTIVE SAMPLING

Selective sampling in the field consists of direct extraction from the environment of items recognizable by the naked eye, usually on the surface of sediments. Sampling for plastic pellets is often selected because their size range (1–6 mm diameter) makes them easily recognizable in the flotsam deposits of sandy beaches. However, there is a great risk of overlooking when microplastics are mixed with other debris or have no characteristic shapes (i.e., irregular, rough, angular). Particular care needs to be taken when selectively sampling them in the field.

6.1.2 BULK SAMPLES

Bulk samples refer to samples where the entire sample volume is taken without reducing it during the sampling process. Bulk samples are most appropriate when microplastics cannot be easily identified visually because:

1. Sediment particles cover them.
2. Their abundance is small, requiring sorting/filtering of large volumes of sediment/water.
3. They are too small to be identified with the naked eye.

6.1.3 VOLUME-REDUCED SAMPLES

Volume-reduced samples in both sediment and seawater samples refer to samples where the volume of the bulk sample is usually reduced during sampling, preserving only that portion of the sample that is of interest for further processing. For sedimentary environments, samples can be sieved directly on the beach or onboard the vessel. In contrast, for seawater samples, volume-reduced samples are usually obtained by filtering large volumes of water with nets. Bulk and volume-reduced samples require further processing in the laboratory.

6.2 AQUATIC SAMPLING OF MICROPLASTICS AND NANOPLASTICS

An aquatic sampling includes freshwater and marine water sampling. Compared to the marine water sampling of microplastics, relatively few studies exist for a freshwater system and estuaries (the transitional zone between freshwater rivers and the marine environment). Most of the time, the collection techniques/ devices are similar. Based on the device used, it is possible to collect different water column layers, from the surface layer (from 0 to 1 m depth; Song et al., 2015) to the bottom layer (Lima et al., 2014). Sampling devices can be divided into three categories:

1. Non-discrete sampling devices
2. Discrete sampling devices
3. Sampling devices of the surface microlayer

6.2.1 NON-DISCRETE SAMPLING DEVICES

6.2.1.1 Nets as Sampling Device

In most of the studies (~76%), nets have been used (usually of 20–800 µm size and 1.5–4.5 m long) for sampling large water volumes from the surface to the bottom layer (Lima et al., 2014; Kang et al., 2015). For this purpose, various nets are used plankton net, manta trawl, catamaran, neuston net, manual net, and continuous net.

The most common type of net used for microplastic sampling in surface waters is the neuston net (as indicated in Figure 6.1), commonly used for the horizontal sampling of both the epineuston (organisms that live in the air on the surface film of the water) and the hyponeuston (organisms that live just beneath the water's surface).

Certainly, surface sampling works best in calm flat waters. Still, in more open waters, where the depth of the net may vary considerably, the use of a more stable manta trawl or catamaran is recommended. The manta trawl has wing-like structures on each side of the net to maintain stability and buoyancy in the water (Figure 6.2).

In contrast to manta trawl, the catamaran has two runners on each side of the net to provide stability and buoyancy as indicated in Figure 6.3 (Crawford and Quinn, 2017).

They are deployed from the ship using structures like a spinnaker, which allows nets to display away from the ship and avoid the turbulence and wave generated by the ship (Palatinus et al., 2019; Green et al., 2018). Usually, the filtered water volume is measured by an attached flow meter (Norén, 2007). Alternatively, it can be calculated as the distance between the starting and end-points (Eriksen et al., 2018). A net aperture consists of a rigid frame that maintains a continuous circular or rectangular

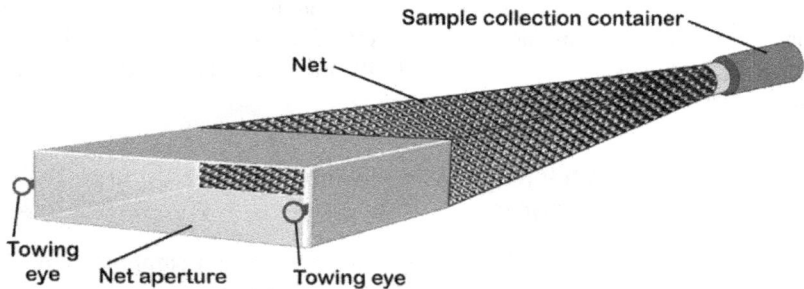

FIGURE 6.1 Sketch of neuston nets used for surface water sampling. (Adopted from Crawford and Quinn, 2017 after permission.)

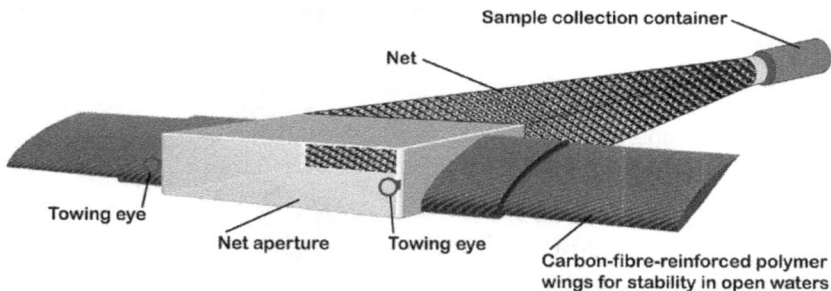

FIGURE 6.2 Sketch of manta trawl used for surface water sampling. (Adopted from Crawford and Quinn, 2017 after permission.)

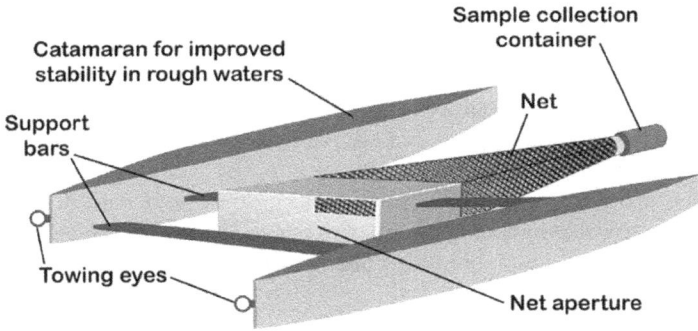

FIGURE 6.3 Sketch of catamaran used for surface water sampling. (Adopted from Crawford and Quinn, 2017 after permission.)

opening at the surface. At the same time, a collecting jar at the end-point helps to concentrate the sample. The net aperture for rectangular openings of neuston nets (sea surface) ranged from 0.03 to $2.0\,m^2$ (van Dolah et al., 1980; Moore et al., 2001). For circular-bongo nets (water column), the net aperture ranged from 0.79 to $1.58\,m^2$ (Carpenter et al., 1972; Doyle et al., 2011). The polyvinyl chloride-built nets contain two wings of resin mixture that helps the aperture to maintain it floating on the water surface or at some specific depth. Notably, a net of 333-μm mesh size is mostly used, while a smaller net (80 μm) can easily get clogged up, making it difficult to do the sampling (Löder et al., 2015), more specifically in a port environment of concentrate suspended particles. Nets can be towed vertically (to sample the water volume from the bottom to surface) or horizontally (to sample the water volume within the surface layer for a period up to 4 h) or obliquely (Collignon et al., 2014; Gorokhova, 2015; Baini et al., 2018; Pan et al., 2019). The flow rate, together with the net aperture area, can be converted into the volume of water filtered (V):

$$V = \pi r^2 d \tag{6.1}$$

Vessels' speed should be calibrated to maintain the flow rate and the quality of collected samples and to avoid turbulence. After sampling, nets need to be rinsed from the outside, preferably with decontaminated water (e.g., Milli-Q water), to avoid sampling contamination. Each sample collected with a net is preserved in a jar until laboratory analysis. The net, once cleaned, can be used for the next sampling. Hence, all sampled microplastic particles are gathered in the collecting jar. Then, the collecting jar is rinsed repeatedly using decontaminated water (e.g., Milli-Q water) and emptied into another jar.

6.2.1.2 Pumping Systems

A pumping system is a less used sampling system (~16%) than the nets (76%). Using various types of pumps (Lusher et al., 2017) allows sampling a large seawater volume from the vessel (Morgana et al., 2018). There are no standard procedures for pumping the seawater; they can be lowered toward the ship's stern (Ng and Obbard,

2006) from a few minutes to hours (Zobkov et al., 2019) and allow sampling from the surface to a depth of 6–100 m (Morgana et al., 2018) under a preferable speed of ship at 1–12 knots (Setälä et al., 2016). The sampled microplastics' dimension depends on the filters/sieves used for the pumping process, allowing the selective size of plastics. A combination of filters and sieves can also be used to divide microplastics into size classifications (Desforges et al., 2014; Cai et al., 2017). Later on, the microplastic mass can be analyzed on filters, while sieves can be rinsed with decontaminated water; whereas, the flowmeter connected with pumping systems can be used to determine the filtration rate and microplastic abundance (Cincinelli et al., 2017).

6.2.2 Discrete Sampling Devices

This type of device is used to sample water at a specific depth, which includes Niskin bottle (Bagaev et al., 2018), rosette (Dai et al., 2018), integrated water sampler (Tamminga et al., 2018). Besides this, bottles, buckets, and steel samplers are also used by dipping manually or through a winch from the ship to the sampling depth in seawater (Bagaev et al., 2018). Thereafter, the collected sample is transferred to a jar and the device is rinsed with decontaminated water to collect microplastics that remain attached. Alternatively, discrete samples can be filtered or sieved directly on board.

6.2.3 Sampling Devices for the Surface Microlayer

Generally, the sampling devices were applied to collect seawater from the first micrometers of the water column using the sieves made of stainless steel of mesh size 2 mm and a rotating drum sampler placed on the water surface (Ng and Obbard, 2006; Song et al., 2015). During sampling, the sieve is placed in contact with the seawater surface, which is retained within the mesh of the sieve due to surface tension and is collected in a stainless-steel plate. This procedure is repeated for a specific amount of time, e.g., 100 times (Song et al., 2015). Water collected in the plate is then transferred to a jar and preserved until laboratory analysis. A rotating drum sampler is placed on the seawater surface at the beginning of sampling. It consists of a glass cylinder that rotates while partially submerged, and it has a hydrophilic clean surface, which, using capillary force, can collect samples from the surface microlayer (from 1 to 1000 μm depths) (Ng and Obbard, 2006; Song et al., 2015).

Overall, the choice of any device to use depends on the available device, aim of the study, type of environment, and matrix that has to be collected. The selected sampling devices can impact the quantity and representativeness of the sampled material. There is no unique, valid protocol for microplastics/nanoplastics sampling in surface waters and each device has merits and associated challenges for sampling, as indicated in Table 6.1 (Campanale et al., 2020b).

6.3 SEDIMENTS SAMPLING OF MICROPLASTICS AND NANOPLASTICS

The distribution of microplastics and nanoplastics on sediments is uneven, largely influenced by their properties and environmental factors, such as winds and currents.

TABLE 6.1

Aquatic Sampling of Microplastics and Nanoplastics with Various Devices and Associated Challenges

Device	Merits	Associated Challenges	Costs$/Time (min)	References
		Non-discrete Sampling Devices		
Manta net	Sampling of large volumes of water; The lateral wings allow the floating of the device and the sampling of the water surface.	Expensive equipment; Requires boat; The lower limit of detection is 333 μm; Clogging problems; Risk of sample contamination; Underestimation of the total buoyant microplastic amounts	~3,500/ 15–240	Cutroneo et al. (2020); Liedermann et al. (2018); Kang et al. (2015); Güven et al. (2017); Anderson et al. (2017); Kooi et al. (2016); Lenaker, et al. (2019); Karlsson et al. (2020)
Neuston net	Sampling of large volumes of water; Widely used (useful for comparing positions)	Expensive equipment; Requires a boat; The lower limit of detection is 333 μm; Clogging problems; Risk of sample contamination; Underestimation of the total buoyant microplastic amounts.	~2,300/ 30	Cutroneo et al. (2020); Collignon et al. (2012); McCormick et al. (2016); Barrows et al. (2017); Syakti et al. (2018)
Plankton net	The lower limit of detection is 100 μm; Sampling of medium volumes of water; Possibility to sample the water column.	Expensive equipment; Requires a boat; Clogging problems; Sampling of lower volumes of water compared to Manta trawl; Risk of sample contamination; Underestimation of the total buoyant microplastic amounts	~2,400/30	Zhang (2017); Cutroneo et al. (2020); Amélineau et al. (2016); Tsang et al. (2017); Dris et al. (2018)

(*Continued*)

TABLE 6.1 (*Continued*)

Aquatic Sampling of Microplastics and Nanoplastics with Various Devices and Associated Challenges

Device	Merits	Associated Challenges	Cost$/Time (min)	References
Microplastic traps	Possibility to sample at several points of the water stream; Possibility to choose mesh dimensions from 100 to 333 μm.	Expensive equipment; It may involve difficulty in anchoring to the riverbed; In the presence of a low flow rate, samples the first 15 cm of water; Risk of contamination.	~1,200/30	Crawford and Quinn (2017)
Autosampler	Well-known and precise volume of filtered water; Minimizes the risk of contamination; Allows a dimensional separation of the particles directly in the field.	Costly equipment; Difficult and heavy to transport and deploy; It may be very fragile; Requires electric energy; Requires a large amount of instrumentation.	10,000–70,000/30	Lenz and Labrenz (2018); Campanale et al. (2020b)
Pumping systems	Allows the user to sample smaller MPs and fiber loss is limited; Well-known and precise volume of filtered water; Allows standardization of sampling	Sampling of a small volume of water; Requires energy to work; Requires boat; It can be challenging to transport and apply. Allows the sampling of a single point; Requires the transport of bulky samples to the lab; Sampling is less representative; Risk of sample contamination.	300–1,000/ 15–180	Setälä et al. (2016); Cutroneo et al. (2020); Tamminga et al. (2018); Zhang et al. (2020); Karlsson et al. (2020); Ng and Obbard (2006)

(Continued)

TABLE 6.1 (*Continued*)

Aquatic Sampling of Microplastics and Nanoplastics with Various Devices and Associated Challenges

Device	Merits	Associated Challenges	Costs$/Time (min)	References
		Discrete Sampling Devices		
Niskin Bottles/ Jars/Bottles/ Buckets/Rosette/ Integrated water sampler/ Ruttner bottles/Friedinger bottles/ Bernatowicz bottles	Relatively quick and straightforward to use; Rosette provides multi-point measurements; Allows sampling at different depths; Allows the user to sample smaller MPs and fiber loss is limited; Well-known and precise volume of filtered water; Allows standardization of sampling.	Requires boat; Rosette can be challenging to transport; Sampling of a small volume of water; It may be very fragile; Requires the transport of bulky samples to the lab; Sampling is less representative; Risk of sample contamination.	Variable/15–30 (300–50,000)	Cutroneo et al. (2020); Dris et al. (2018); Bagaev et al. (2018); Dai et al. (2018); Tamminga et al. (2018); Covernton et al. (2019); Courtene-Jones et al. (2017)
		Devices for Surface Microlayer		
Stainless steel sieves/Rotating Drum Sampler	Does not require specialized equipment; Quick and straightforward to use; Well-known and precise volume of filtered water; Allows choice of mesh size; Allows a dimensional separation of the particles directly in the field	Sampling of medium/low volumes of water; Requires the transport of a significant volume of water to the lab; Manual transfer of water with buckets; Potential contamination by the apparatus	From 50 to variable/ Depending on mesh size	Cutroneo et al. (2020); Ng and Obbard (2006); Song et al. (2015)

Among the studies that determined the microplastic densities in the sedimentary environment, there is a lack of uniformity that deals with microplastic accumulation across the beach zone. In most of the studies, the selective sediment samples for plastic pellets and fragments were taken with tweezers (Ashton et al., 2010) tablespoons (Cooper and Corcoran, 2010) or picked up by hand (Mato et al., 2001; Turner and Holmes, 2011). While studies sampled at the high tide line, using different approaches:

1. Sampling a linear extension along the strandline with a spoon and/or a trowel
2. Sampling an areal extension using quadrats
3. Sampling different depth strata using corers

Most studies do not report exact sampling procedures, but few studies from the sublittoral zone sampled microplastics with Ekman, and Van Veen grab samplers (Browne et al., 2011; Thompson et al., 2004; Claessens et al., 2011). Sampling units were directly related to the sampling instrument used. Studies that sampled a specific areal extension (from 0.0079 to 5 m²) employed quadrats and corers. Other sampling units were weight (from 0.15 to 10 kg) and volume of sediment (from 0.1 to 8 L). Samples were taken to variable depths below the sediment surface. Reported sampling depths ranged from 0 to 32 cm. Most studies sampled a single depth layer within the top 5 cm of sediment (Martins and Sobral, 2011; Cooper and Corcoran, 2010). Some studies sampled a second layer at 10 cm depth, in addition to the sediment surface (Ng and Obbard, 2006; Corcoran et al., 2009).

Few studies followed a stratified sampling scheme using a corer down to a depth of 25 cm, separating the core into five layers, each of which had a thickness of 5 cm (Carson et al., 2011), and to a depth of 28 and 32 cm, with four sediment layers of 7 and 8 cm, respectively (Claessens et al., 2011). Given that beaches and subtidal coastal habitats are dynamic systems with continuous and seasonal sediment erosion (Rusch et al., 2000; Huettel and Rusch, 2000), microplastics may become buried in sediment during periods of accretion. Furthermore, beaches filter and retain particulate organic matter over a range of depths. Sediments between 0 and 5 cm in depth are characterized by steep gradients and strong seasonal variation of more fine-grained particles and particulate organic matter (Rusch et al., 2000). In permeable sands, microplastics might accumulate in similar ways as sediment particles and particulate organic matter, resulting in microplastics being trapped in deeper sediment layers so this should be examined with stratified samples using cores (Carson et al., 2011; Martins and Sobral, 2011).

Sample weight (25–3,000 g) or volume (0.05–1.2 L) largely varies between studies, potentially affecting representativeness. The National Oceanic and Atmospheric Administration recommends the use of 400 g (w.w.) per replicate, followed by drying and weighing to adjust the results, while the EU Marine Strategy Framework Directive technical subgroup recommends using at least five replicates of the top 5 cm of sediment (Prata et al., 2019). So here, a standardized protocol is yet needed to be established. Various sediment sampling devices are sketched in Figure 6.4.

FIGURE 6.4   Sketch of various sediment sampling devices. (Adopted from Campanale et al., 2020b; Vianello et al., 2013 and Castañeda et al., 2014.)

6.4   ATMOSPHERIC SAMPLING OF MICROPLASTICS AND NANOPLASTICS

For an atmospheric sample collection, a range of wet/dry deposition is carried out using a passive collector (total deposition) (Allen et al., 2019, Cai et al., 2017; Dris et al., 2017; Klein and Fischer, 2019). Although the sampling via sweeping, vacuum, and the active pump is easier to perform, the data comparison is quite difficult for determining the relative quantity of sampled air or whether the collected microplastics belong exclusively to the atmospheric deposition (Liu et al., 2019).

The Norwegian Institute for Air Research has recently developed a standard system for passive atmospheric deposition of microplastics. Besides the plastic-free standardized method, the ease of use, zero-power requirement, and the volume of blow-by sample is a known volume that allows for comparison to other deposited material in addition to other plastic studies. The active pumped air sampling is another standard protocol for site-specific meteorological conditions and known terrestrial or ocean surface conditions. It is an established method for atmospheric pollution monitoring (Dommergue et al., 2019; Hayward et al., 2010). Dris et al. (2017) used active air pumped sampling methodology to enable a known volume of indoor air to be sampled (filtered).

ACKNOWLEDGMENTS

This work was supported by a grant from the National Institute of Environment Research (NIER), funded by the Ministry of Environment (MOE) of the Republic of Korea (NIER-RP2020-249) and by the National Research Foundation of Korea (NRF) grant funded by the Korea government (MSIT) (No. NRF-2020R1A2C1013851).

REFERENCES

Anderson, P. J., Warrack, S., Langen, V., Challis, J. K., Hanson, M. L. & Rennie, M. D. (2017) Microplastic contamination in lake Winnipeg, Canada. *Environmental Pollution*, 225, 223–231.

Allen, S., Allen, D., Phoenix, V. R., Le Roux, G., Jiménez, P. D., Simonneau, A. & Galop, D. (2019) Atmospheric transport and deposition of microplastics in a remote mountain catchment. *Nature Geoscience*, 12(5), 339–344.

Ashton, K., Holmes, L. & Turner, A. (2010) Association of metals with plastic production pellets in the marine environment. *Marine Pollution Bulletin*, 60(11), 2050–2055.

Amélineau, F., Bonnet, D., Heitz, O., Mortreux, V., Harding, A. M., Karnovsky, N. & Gremillet, D. (2016) Microplastic pollution in the Greenland Sea: background levels and selective contamination of planktivorous diving seabirds. *Environmental Pollution*, 219, 1131–1139.

Bagaev, A., Khatmullina, L. & Chubarenko, I. (2018) Anthropogenic microlitter in the Baltic Sea water column. *Marine Pollution Bulletin*, 129(2), 918–923.

Baini, M., Fossi, M. C., Galli, M., Caliani, I., Campani, T., Finoia, M. G. & Panti, C. (2018) Abundance and characterization of microplastics in the coastal waters of Tuscany (Italy): the application of the MSFD monitoring protocol in the Mediterranean Sea. *Marine Pollution Bulletin*, 133, 543–552.

Barrows, A. P., Neumann, C. A., Berger, M. L. & Shaw, S. D. (2017) Grab vs. neuston tow net: a microplastic sampling performance comparison and possible advances in the field. *Analytical Methods*, 9(9), 1446–1453.

Browne, M. A., Crump, P., Niven, S. J., Teuten, E., Tonkin, A., Galloway, T. & Thompson, R. (2011) Accumulation of microplastic on shorelines woldwide: sources and sinks. *Environmental Science & Technology*, 45(21), 9175–9179.

Cai, L., Wang, J., Peng, J., Tan, Z., Zhan, Z., Tan, X. & Chen, Q. (2017) Characteristic of microplastics in the atmospheric fallout from Dongguan city, China: preliminary research and first evidence. *Environmental Science and Pollution Research*, 24(32), 24928–24935.

Carpenter, E. J., Anderson, S. J., Harvey, G. R., Miklas, H. P. & Peck, B. B. (1972) Polystyrene spherules in coastal waters. *Science*, 178(4062), 749–750.

Carson, H. S., Colbert, S. L., Kaylor, M. J. & McDermid, K. J. (2011) Small plastic debris changes water movement and heat transfer through beach sediments. *Marine Pollution Bulletin*, 62(8), 1708–1713.

Castaneda, R. A., Avlijas, S, & Simard, M. A. (2014) Microplastic pollution in St. Lawrence River sediments. *Canadian Journal of Fisheries and Aquatic Sciences*, 71, 1767–1771.

Cincinelli, A., Scopetani, C., Chelazzi, D., Lombardini, E., Martellini, T., Katsoyiannis, A. & Corsolini, S. (2017) Microplastic in the surface waters of the Ross Sea (Antarctica): occurrence, distribution and characterization by FTIR. *Chemosphere*, 175, 391–400.

Claessens, M., De Meester, S., Van Landuyt, L., De Clerck, K. & Janssen, C. R. (2011) Occurrence and distribution of microplastics in marine sediments along the Belgian coast. *Marine Pollution Bulletin*, 62(10), 2199–2204.

Campanale, C., De Palma, C. P., Bollino, B., Massarelli, C. & Uricchio, V. F. (2020a) Innovativo sistema di campionamento automatizzato per il monitoraggio di microplastiche in ambienti fluviali. In *Proceedings of the RemTech Expo–Hub tecnologica Campania*, Napoli, Italy.

Campanale, C., Savino, I., Pojar, I., Massarelli, C., Uricchio, V. F. (2020b). A practical overview of methodologies for sampling and analysis of microplastics in riverine environments. *Sustainability*, 12, 6755; doi:10.3390/su12176755.

Collignon, A., Hecq, J. H., Galgani, F., Collard, F. & Goffart, A. (2014) Annual variation in neustonic micro-and meso-plastic particles and zooplankton in the Bay of Calvi (Mediterranean–Corsica). *Marine Pollution Bulletin*, 79(1–2), 293–298.

Collignon, A., Hecq, J. H., Glagani, F., Voisin, P., Collard, F. & Goffart, A. (2012) Neustonic microplastic and zooplankton in the North Western Mediterranean Sea. *Marine Pollution Bulletin*, 64(4), 861–864.

Cooper, D. A. & Corcoran, P. L. (2010) Effects of mechanical and chemical processes on the degradation of plastic beach debris on the island of Kauai, Hawaii. *Marine Pollution Bulletin*, 60(5), 650–654.

Corcoran, P. L., Biesinger, M. C. & Grifi, M. (2009) Plastics and beaches: a degrading relationship. *Marine Pollution Bulletin*, 58(1), 80–84.

Covernton, G. A., Pearce, C. M., Gurney-Smith, H. J., Chastain, S. G., Ross, P. S., Dower, J. F. & Dudas, S. E. (2019) Size and shape matter: a preliminary analysis of microplastic sampling technique in seawater studies with implications for ecological risk assessment. *Science of the Total Environment*, 667, 124–132.

Crawford, C. B. & Quinn, B. (2017) *Microplastic Collection Techniques. Microplastic Pollutants*; Elsevier Science: Amsterdam, The Netherland, pp. 179–202.

Cutroneo, L., Reboa, A., Besio, G., Borgogno, F., Canesi, L., Canuto, S. & Capello, M. (2020) Microplastics in seawater: sampling strategies, laboratory methodologies, and identification techniques applied to port environment. *Environmental Science and Pollution Research*, 27(9), 8938–8952.

Courtene-Jones, W., Quinn, B., Gary, S. F., Mogg, A. O. & Narayanaswamy, B. E. (2017) Microplastic pollution identified in deep-sea water and ingested by benthic invertebrates in the Rockall Trough, North Atlantic Ocean. *Environmental Pollution*, 231, 271–280.

Dai, Z., Zhang, H., Zhou, Q., Tian, Y., Chen, T., Tu, C. & Luo, Y. (2018) Occurrence of microplastics in the water column and sediment in an inland sea affected by intensive anthropogenic activities. *Environmental Pollution*, 242, 1557–1565.

Desforges, J. P. W., Galbraith, M., Dangerfield, N. & Ross, P. S. (2014) Widespread distribution of microplastics in subsurface seawater in the NE Pacific Ocean. *Marine Pollution Bulletin*, 79(1–2), 94–99.

Dommergue, A., Amato, P., Tignat-Perrier, R., Magand, O., Thollot, A., Joly, M., Bouvier, L., Sellegri, K., Vogel, T., Sonke, J.E., Jaffrezo, J.L., Andrade, M., Moreno, I., Labuschagne, C., Martin, L., Zhang, Q. & Larose, C. (2019) Methods to investigate the global atmospheric microbiome. *Frontiers in Microbiology*, 10, 243.

Doyle, M. J., Watson, W., Bowlin, N. M. & Sheavly, S. B. (2011) Plastic particles in coastal pelagic ecosystems of the Northeast Pacific Ocean. *Marine Environmental Research*, 71(1), 41–52.

Dris, R., Gasperi, J., Mirande, C., Mandin, C., Guerrouache, M., Langlois, V. & Tassin, B. (2017) A first overview of textile fibers, including microplastics, in indoor and outdoor environments. *Environmental Pollution*, 221, 453–458.

Dris, R., Gasperi, J. & Tassin, B. (2018) Sources and fate of microplastics in urban areas: a focus on Paris megacity. In; Wagner, M., Lambert, S., (Eds.). *Freshwater Microplastics. The Handbook of Environmental Chemistry*. Springer: Cham, Switzerland. pp. 69–83.

EC. (2013) *GREEN PAPER: On a European Strategy on Plastic Waste in the Environment*. EU (European Commission): Brussels, Belgium.

Eriksen, M., Mason, S., Wilson, S., Box, C., and Amato, S. (2013) Microplastic pollution in the surface waters of the Laurentian Great Lakes. *Marine Pollution Bulletin*, 77, 177–182. doi: 10.1016/j.marpolbul.2013.10.007.

Gorokhova, E. (2015) Screening for microplastic particles in plankton samples: how to integrate marine litter assessment into existing monitoring programs? *Marine Pollution Bulletin*, 99(1–2), 271–275.

Green, D. S., Kregting, L., Boots, B., Blockley, D. J., Brickle, P., Da Costa, M. & Crowley, Q. (2018) A comparison of sampling methods for seawater microplastics and a first report of the microplastic litter in coastal waters of Ascension and Falkland Islands. *Marine Pollution Bulletin*, 137, 695–701.

Güven, O., Gökdağ, K., Jovanović, B. & Kıdeyş, A. E. (2017) Microplastic litter composition of the Turkish territorial waters of the Mediterranean Sea, and its occurrence in the gastrointestinal tract of fish. *Environmental Pollution*, 223, 286–294.

Hayward, S. J., Gouin, T. & Wania, F. (2010) Comparison of four active and passive sampling techniques for pesticides in air. *Environmental Science & Technology*, 44(9), 3410–3416.

Hidalgo-Ruz, V., Gutow, L., Thompson, R. C. & Thiel, M. (2012) Microplastics in the marine environment: a review of the methods used for identification and quantification. *Environmental Science & Technology*, 46(6), 3060–3075.

Hidalgo-Ruz, V. & Thiel, M. (2013) Distribution and abundance of small plastic debris on beaches in the SE Pacific (Chile): a study supported by a citizen science project. *Marine Environmental Research*, 87–88, 12–18.

Huettel, M. & Rusch, A. (2000) Transport and degradation of phytoplankton in permeable sediment. *Limnology and Oceanography*, 45(3), 534–549.

Kang, J. H., Kwon, O. Y., Lee, K. W., Song, Y. K. & Shim, W. J. (2015) Marine neustonic microplastics around the southeastern coast of Korea. *Marine Pollution Bulletin*, 96(1–2), 304–312.

Karlsson, T. M., Kärrman, A., Rotander, A. & Hassellöv, M. (2020) Comparison between manta trawl and in situ pump filtration methods, and guidance for visual identification of microplastics in surface waters. *Environmental Science and Pollution Research*, 27(5), 5559–5571.

Klein, M. & Fischer, E. K. (2019) Microplastic abundance in atmospheric deposition within the Metropolitan area of Hamburg, Germany. *Science of the Total Environment*, 685, 96–103.

Kooi, M., Reisser, J., Slat, B., Ferrari, F. F., Schmid, M. S., Cunsolo, S. & Koelmans, A. A. (2016) The effect of particle properties on the depth profile of buoyant plastics in the ocean. *Scientific Reports*, 6(1), 1–10.

Lenz, R. & Labrenz, M. (2018) Small microplastic sampling in water: development of an encapsulated filtration device. *Water*, 10(8), 1055.

Lenaker, P. L., Baldwin, A. K., Corsi, S. R., Mason, S. A., Reneau, P. C. & Scott, J. W. (2019) Vertical distribution of microplastics in the water column and surficial sediment from the Milwaukee River Basin to Lake Michigan. *Environmental Science & Technology*, 53(21), 12227–12237.

Lima, A. R. A., Costa, M. F. & Barletta, M. (2014) Distribution patterns of microplastics within the plankton of a tropical estuary. *Environmental Research*, 132, 146–155.

Liu, C., Li, J., Zhang, Y., Wang, L., Deng, J., Gao, Y. & Sun, H. (2019) Widespread distribution of PET and PC microplastics in dust in urban China and their estimated human exposure. *Environment International*, 128, 116–124.

Liedermann, M., Gmeiner, P., Pessenlehner, S., Haimann, M., Hohenblum, P. & Habersack, H. (2018) A methodology for measuring microplastic transport in large or medium rivers. *Water*, 10(4), 414.

Löder, M. G. J., Kuczera, M., Mintenig, S., Lorenz, C. & Gerdts, G. (2015) Focal plane array detector-based micro-Fourier-transform infrared imaging for the analysis of microplastics in environmental samples. *Environmental Chemistry*, 12(5), 563–581.

Lusher, A. L., Welden, N. A., Sobral, P. & Cole, M. (2017) Sampling, isolating and identifying microplastics ingested by fish and invertebrates. *Anal Methods*, 9, 1346–1360.

Martins, J. & Sobral, P. (2011) Plastic marine debris on the Portuguese coastline: a matter of size? *Marine pollution Bulletin*, 62(12), 2649–2653.

Mato, Y., Isobe, T., Takada, H., Kanehiro, H., Ohtake, C. & Kaminuma, T. (2001) Plastic resin pellets as a transport medium for toxic chemicals in the marine environment. *Environmental Science & Technology*, 35(2), 318–324.

McCormick, A. R., Hoellein, T. J., London, M. G., Hittie, J., Scott, J. W. & Kelly, J. J. (2016) Microplastic in surface waters of urban rivers: concentration, sources, and associated bacterial assemblages. *Ecosphere*, 7(11), e01556.

Moore, C. J., Moore, S. L., Leecaster, M. K. & Weisberg, S. B. (2001) A comparison of plastic and plankton in the North Pacific central gyre. *Marine Pollution Bulletin*, 42(12), 1297–1300.

Morgana, S., Ghigliotti, L., Estévez-Calvar, N., Stifanese, R., Wieckzorek, A., Doyle, T. & Garaventa, F. (2018) Microplastics in the Arctic: a case study with sub-surface water and fish samples off Northeast Greenland. *Environmental Pollution*, 242, 1078–1086.

Ng, K. L. & Obbard, J. P. (2006) Prevalence of microplastics in Singapore's coastal marine environment. *Marine Pollution Bulletin*, 52(7), 761–767.

Norén, F. (2007) Small plastic particles in coastal Swedish waters. *Kimo Sweden*, 11, 1–11.

Palatinus, A., Viršek, M. K. & Kaberi, E. (2019) DeFishGear protocols for sea surface and beach sediment sampling and sample analysis, 27.

Pan, Z., Guo, H., Chen, H., Wang, S., Sun, X., Zou, Q. & Huang, J. (2019) Microplastics in the Northwestern Pacific: abundance, distribution, and characteristics. *Science of the Total Environment*, 650, 1913–1922.

Prata, J. C., da Costa, J. P., Duarte, A. C. & Rocha-Santos, T. (2019) Methods for sampling and detection of microplastics in water and sediment: a critical review. *TrAC Trends in Analytical Chemistry*, 110, 150–159.

Rusch, A., Huettel, M. & Forster, S. (2000) Particulate organic matter in permeable marine sands—dynamics in time and depth. *Estuarine, Coastal and Shelf Science*, 51(4), 399–414.

Setälä, O., Magnusson, K., Lehtiniemi, M. & Norén, F. (2016) Distribution and abundance of surface water microlitter in the Baltic Sea: a comparison of two sampling methods. *Marine Pollution Bulletin*, 110(1), 177–183.

Song, Y. K., Hong, S. H., Jang, M., Han, G. M., Rani, M., Lee, J. & Shim, W. J. (2015) A comparison of microscopic and spectroscopic identification methods for analysis of microplastics in environmental samples. *Marine Pollution Bulletin*, 93(1–2), 202–209.

Syakti, A. D., Hidayati, N. V., Jaya, Y. V., Siregar, S. H., Yude, R., Asia, L. & Doumenq, P. (2018) Simultaneous grading of microplastic size sampling in the Small Islands of Bintan water, Indonesia. *Marine Pollution Bulletin*, 137, 593–600.

Tamminga, M., Hengstmann, E. & Fischer, E. K. (2018) Microplastic analysis in the South Funen Archipelago, Baltic Sea, implementing manta trawling and bulk sampling. *Marine Pollution Bulletin*, 128, 601–608.

Thompson, R. C., Olsen, Y., Mitchell, R. P., Davis, A., Rowland, S. J., John, A. W. & Russell, A. E. (2004) Lost at sea: where is all the plastic? *Science*, 304(5672), 838–838.

Tsang, Y. Y., Mak, C. W., Liebich, C., Lam, S. W., Sze, E. T. & Chan, K. M. (2017) Microplastic pollution in the marine waters and sediments of Hong Kong. *Marine Pollution Bulletin*, 115(1–2), 20–28.

Turner, A. & Holmes, L. (2011) Occurrence, distribution and characteristics of beached plastic production pellets on the island of Malta (central Mediterranean). *Marine Pollution Bulletin*, 62(2), 377–381.

Vesilind, P. (2003) *Wastewater Treatment Plant Design*. Vol. 2: IWA Publishing. ISBN 1-84339-029-8.

van Dolah, R. F., Burrell Jr, V. G. & West, S. B. (1980) The distribution of pelagic tars and plastics in the South Atlantic Bight. *Marine Pollution Bulletin*, 11(12), 352–356.

Vianello, A., Boldrin, A., Guerriero, P., Moschino, V., Rella, R., Sturaro, A., Da Ros, L. (2013) Microplastic particles in sediments of Lagoon of Venice, Italy: First observations on occurrence, spatial patterns and identification. *Estuarine, Coastal and Shelf Science*, 130, 54–61.

Wheatley, L., Levendis, Y. A., Vouros, P. (1993) Exploratory-study on the combustion and Pah emissions of selected municipal waste plastics. *Environmental Science & Technology*, 27(13), 2885–2895.

Zhang, K., Gong, W., Lv, J., Xiong, X., Wu, C. (2015) Accumulation of floating microplastics behind the Three Gorges Dam. *Environmental Pollution*, 204, 117–123.

Zhang, H. (2017) Transport of microplastics in coastal seas. *Estuarine, Coastal and Shelf Science*, 199, 74–86.

Zhang, L., Liu, J., Xie, Y., Zhong, S., Yang, B., Lu, D. & Zhong, Q. (2020) Distribution of microplastics in surface water and sediments of Qin river in Beibu Gulf, China. *Science of the Total Environment*, 708, 135176.

Zobkov, M. B., Esiukova, E. E., Zyubin, A. Y. & Samusev, I. G. (2019) Microplastic content variation in water column: the observations employing a novel sampling tool in stratified Baltic Sea. *Marine Pollution Bulletin*, 138, 193–205.

7 Separation of Microplastics and Nanoplastics from Various Environments and Associated Challenges

Hyunjung Kim, Sadia Ilyas, Allan Gomez-Flores
Hanyang University

Humma Akram Cheema
University of Agriculture Faisalabad (UAF)

CONTENTS

DOI: 10.1201/9781003200628-7

Once the sample collection process ends, different sample preservation and separation methods can be used for microplastics and nanoplastics. To preserve the biological component of samples for laboratory analysis, preservation methods mainly involve the use of 4% formalin, which is added to the sample before it is sealed for transfer to the laboratory (van der Hal et al. 2017; Figueiredo and Vianna 2018). If MPs are the only parameters of interest in a study, ethanol or sample refrigeration can be used as a preservative (Castillo et al. 2016; Viršek et al. 2016).

Collected samples must undergo one or more separation processes to isolate the MP within the sample. This ensures that the microplastics can be quantified by counting or weighing and positively identified. It is noteworthy that MPs can be directly identified and classified, without separation processes, only when it remains already collected on a filter after sampling. Separation methods can be classified into the following: density separation, filtration, and sieving (Hidalgo-Ruz et al. 2012). Visual sorting of samples can be performed prior to filtration or sieving (for aquatic, sediment, atmospheric, and biological samples if required) to remove plastics with sizes larger than 5 mm, which are visible with the naked eye (Maes et al. 2017), or it can also be performed after sieving by removing the plastics retained on the sieve (Viršek et al. 2016). Moreover, different kinds of sample digestion (acidic, enzymatic, alkaline, and oxidative) can be performed to separate microplastics or nanoplastics from biological matter (Miller et al. 2017).

It is noteworthy that all preparation techniques are interconnected and more than one can be used for a particular type of sample depending upon the specific goal. Visual separation, as a prerequisite step for all sample types still have inherent limitations, as indicated by Eriksen et al. (2013). They described the misidentification of approximately 20% of the particles initially identified as microplastics by visual observation, which were subsequently identified as aluminum silicate from coal ash using scanning electron microscopy. In other works, 32% of visually counted microplastic particles below 100 μm were not confirmed as microplastics after micro-Raman application (Enders et al. 2015) and up to 70% of particles were erroneously identified as microplastics after FTIR analysis (Hidalgo-Ruz et al. 2012). Concerning nanoplastics, there are still no established protocols for their separation, identification, and quantification in complex samples.

7.1 SEPARATION OF MICROPLASTICS FROM WATER SAMPLES BY VARIOUS TECHNIQUES AND ASSOCIATED CHALLENGES

Figure 7.1 depicts schematics of different steps of laboratory methodologies applied to separate microplastics from the aquatic system. Most water samples taken from the environment for microplastic analysis are subjected to some form of volume reduction at the time the sample was taken. This is usually undertaken by filtration with a net or sieving. In other cases, the collection of bulk water samples may have been undertaken. However, once the samples arrive at the laboratory, microplastics are routinely removed from both "volume reduced" and "bulk water" samples by the standard laboratory techniques of sieving and filtration.

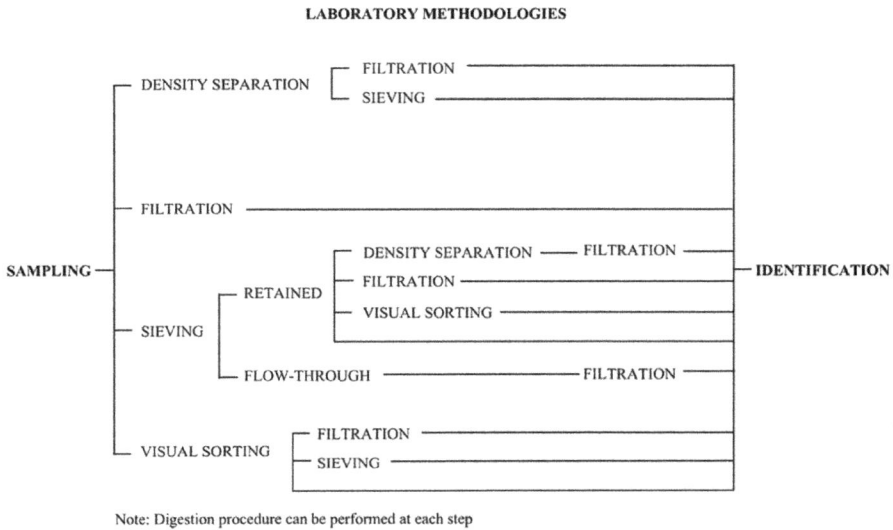

FIGURE 7.1 Scheme of different steps of laboratory methodologies applied to separate microplastics from the aquatic system. (Adopted from Cutroneo et al., 2020 under a Creative Commons Attribution 4.0 International License.)

7.1.1 FILTRATION AS SEPARATION TECHNIQUE

Filtration is an effective mechano-physical technique to solid–liquid separation through a medium that allows only passing the liquid called filtrate. The solids retaining capacity depends upon the pore size of the filter medium (cloths/papers) (Hidalgo-Ruz and Thiel 2013; Titow 1986) while using a gravity or vacuum filtration unit (Figure 7.2).

Various types of filters are used for microplastics separation from seawater, which include nylon (Tang et al. 2018), polycarbonate (Norén 2007), glass fiber (Pan et al. 2019), polyamide (Enders et al. 2015), cellulose acetate (Castro et al. 2016), cellulose nitrate (Dubaish and Liebezeit 2013), mixed cellulose ester (Desforges et al. 2014), and Anopore inorganic membrane filters (Saliu et al. 2018). The filters of pore sizes range between 0.2 and 300 μm (Setälä et al. 2016; Syakti et al. 2018), while a filter of diameter 45−47 mm (Norén 2007; Zhu et al. 2019) is the most frequently used, such as VWR Grade 310. Samples can be directly filtered/sieved, which helps to retain microplastic particles, thereafter, removing the microplastics using decontaminated water (Lusher et al. 2017). Further, the microplastic particles retained by sieving can be directly sent to visual analysis, while those that pass through the sieve are filtered (Erni-Cassola et al. 2017). Else, the original sample can be settled to deposit the inorganic and biological materials at the bottom followed by filtration of supernatant while precipitating the sieved portion (Castro et al. 2016). Finally, filter is rinsed with pure water before preserving to avoid salt formation on the dry filters, while the remaining solution can be dried at 40°C–70°C (Barrows et al. 2017; Hall et al. 2015) as most of the plastics melt at ≥100°C (Sigma Aldrich 2019). Thus obtained dried

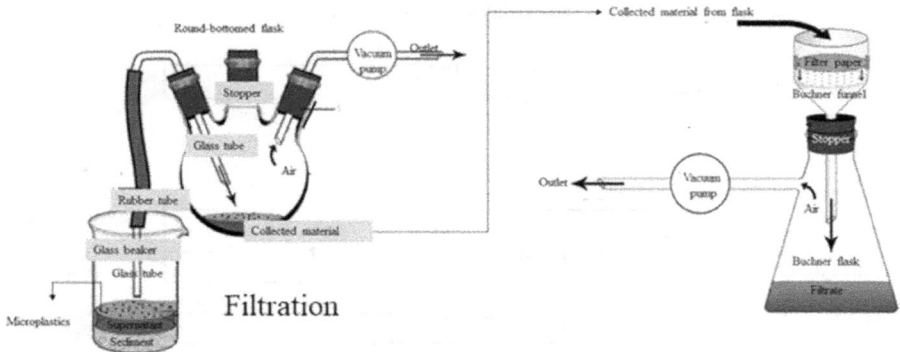

FIGURE 7.2 General Sketch of vacuum filtration for separation of various plastics. (Significantly modified from Crawford and Quinn, 2017).

filters are weighed to get the microplastic mass (Song et al. 2015) or a digestion can also be done (Abayomi et al. 2017).

It is notable that the contaminated sample contains particulate matter or debris which can clog the filter paper and reduce the filtration rate. This can be controlled by reducing the sample volume to be filtered and by adding a pre-filtration cleaning step that includes settling, density separation, flocculation, coagulation, etc.

7.1.2 SIEVING AS SEPARATION TECHNIQUE

A general sketch of sieving system for the separation of various plastics is presented in Figure 7.3.

The sieve physically captures the microplastics and allows the water to be lost from the sample. The size of the mesh used in the sieve depends upon the size of the microplastics that are desired to be collected. While the mesh size used in most studies ranged from 38 μm to 4.75 mm. (Crawford and Quinn, 2017). Notably, the sieving mesh size must not exceed 5 mm, else the collected size of plastics would be outside the size range of microplastics. The most commonly used method to sieve water and sediment samples for microplastics is multi-tier sieving. This involves separating material of different sizes by passing the sample through a series of sieves with a decreasing mesh size (Lima et al. 2014; Masura et al. 2015; Suaria et al. 2016; Kroon et al. 2018). This approach is also advantageous for tertiary treated municipal effluents using mesh sizes between 400 and 20 μm (Maes et al. 2017). The retained microplastics by the sieve undergo visual sorting, whereas the sieve flow is sent to filtration (Kang et al. 2015; Erni-Cassola et al. 2017). Thus obtained microplastics after filtration can be dried as described in the above section before weighing the mass. Alternatively, microplastics after being concentrated can be picked out of the sieve using tweezers but this may be prone to greater error. The technique is unsuccessful for complex raw effluent with high organic contents as the sieves became rapidly clogged.

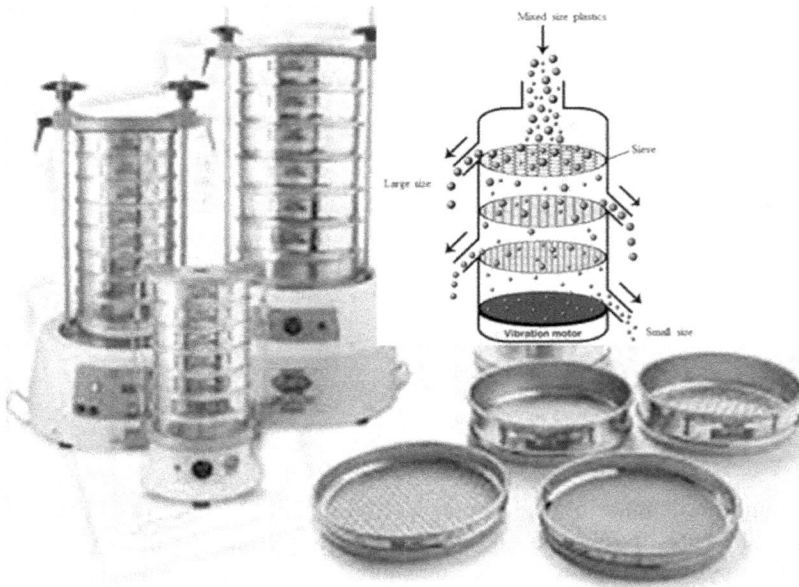

FIGURE 7.3 General Sketch of sieving system for separation of various plastics.

7.2 SEPARATION OF MICROPLASTICS FROM SEDIMENT SAMPLES BY VARIOUS TECHNIQUES AND ASSOCIATED CHALLENGES

The separation of microplastics from sediment is mainly influenced by the physical characteristics, viz. size, shape, and density of both sediment and the microplastics of interest. The grain sizes applicable to common microplastic research are listed in Table 7.1 (Crawford and Quinn 2017). Due to the listed characteristics, it is not easy to separate small microplastics from sediment consisting of silt, clay, or colloid. Indeed, the finer the microplastics and sediment, the more difficult is their separation. Furthermore, the fiber is harder to separate than spherical shapes of the microplastics. Nevertheless, sieving and filtration of finer sediments is an appropriate technique to extract microplastics.

7.2.1 DENSITY-BASED SEPARATION TECHNIQUE

In this process, the actuality of plastic materials is exploited to separate the materials differing in their densities from each other like the sediments and the microplastics. When the liquid mixture (or, say, slurry) of intermediate density is placed, the lower density material floats at the upper layer while a greater density material sinks at the bottom. The floating microplastics are thus collected by decantation of upper layer and leaving sediments at the bottom portion. Usually, the density of salt liquid (e.g., NaCl solution of density 1.202 g cm^{-3}) can be adjusted to allow plastic materials floating on the surface and facilitate a faster separation (Hidalgo-Ruz and Thiel 2013). Importantly, unnecessary atmospheric contact of the sample must be avoided

TABLE 7.1

Common Sediment Grain Sizes

Aggregate	Grain Diameter
Fine gravel	8–4 mm
Very fine gravel	4–2 mm
Very coarse sand	2–1 mm
Coarse sand	1 mm–500 μm
Medium sand	500–250 μm
Fine sand	250–125 μm
Very fine sand	125–62.5 μm
Silt	62.5–3.9 μm
Clay	3.9–1 μm
Colloid	<1 μm

by keeping the number of steps limited to minimum possible steps; however, the salt solutions use, the sample mixing time (ranging from 30 s to 2 h), the settling time (ranging from 2 min to 6 h) and the number of replicates can vary as required (Hidalgo-Ruz and Thiel 2013). For example, if a low-density solution is desired to separate low-density plastics, the saturated NaCl solution should be replaced by ethanol diluted in distilled water. Further, by changing ethanol-to-water ratio, the solutions with a lower density than that of water can be prepared. The majority of plastic materials are seldom composed of the pure polymer and are normally blended with glass fiber, or other materials, as well as being plasticized and impact modified, which results in density change of the plastics that often gets increased. The minimum and maximum density values for common plastics are listed in Table 7.2.

The dense microplastics, such as polyvinyl chloride (PVC) and polyethylene terephthalate (PET), cannot be separated from sediments if using a saturated NaCl solution. Use of heavy liquids, like zinc chloride (Lima et al. 2014; Imhof et al. 2013), sodium polytungstate (Corcoran et al. 2015; Zhao et al. 2014), sodium iodide (Cole and Sherrington 2016; Olencycz et al. 2015) and lithium metatungstate solutions can be suitable one. Notably, solutions with a density of more than 1.4 g cm^{-3} are suggested to ensure the efficient separation of all plastic materials from the sediment. Mostly a high-density solution can extract plastics more effectively (Table 7.2).

A repetitive density separation can achieve effective extraction of microplastics. For example, with a sodium chloride (NaCl) solution, the extraction efficiency of polyethylene microplastics increased from 61% for the first extraction to 83% for the second and reaching up to 93% for the third contact (Titow, 1986), which can be improved up to 99% (Olencycz et al. 2015) and 100% (Cole and Sherrington 2016) using NaI and ZnCl$_2$ solution, respectively. However, the fibers' extraction from sediments samples and the recovery of 40–309 μm mini microplastics from sediments are found to be difficult (Imhof et al. 2013). While heavy liquids assist in better recoveries of microplastics from sediment samples, they are generally very expensive to procure and some are even considered to be toxic to the environment. Table 7.3 enlists various reagents used for density separation along with their cost, density, solubility, and toxicity.

TABLE 7.2
List of Some Common Plastics with Density Ranges and Common
High-Density Solutions to Separate Sediment Microplastics

Plastics/Abbreviation	Density (g cm^{-3})
Poly(methyl methacrylate)/PMA	1.17–1.20
Poly(methyl methacrylate) (high heat)/PMA	1.15–1.25
Poly(methyl methacrylate) (impact modified)/PMA	1.10–1.20
Polytetrafluoroethylene/PTFE	2.10–2.20
Polytetrafluoroethylene (25% glass fiber)/PTFE	2.20–2.30
Polypropylene (impact modified)/PP	0.88–0.91
Polypropylene (homopolymer)/PP	0.90–0.91
Polypropylene (copolymer)/PP	0.90–0.91
Polypropylene (10%–20% glass fiber)/PP	0.97–1.05
Polypropylene (10%–40% mineral filled)/PP	0.97–1.25
Polypropylene (10%–40% talc)/PP	0.97–1.25
Polypropylene (30%–40% glass fiber)/PP	1.10–1.23
Low-density polyethylene/LDPE	0.92–0.94
Linear low-density polyethylene/LLDPE	0.92–0.95
High-density polyethylene/HDPE	0.94–0.97
Polystyrene (crystal)/PS	1.04–1.05
Polystyrene (high heat)/PS	1.04–1.05
Acrylonitrile butadiene styrene/ABS	1.03–1.21
Acrylonitrile butadiene styrene (flame retardant)/ABS	1.15–1.20
Acrylonitrile butadiene styrene (high impact)/ABS	1.00–1.10
Acrylonitrile butadiene styrene (high heat)/ABS	1.00–1.15
Nylon 6,6 (impact modified)/PA	1.05–1.10
Nylon 6/PA	1.12–1.14
Nylon 6,6/PA	1.13–1.15
Nylon 6,6 (30% glass fiber)/PA	1.37–1.38
Nylon 6,6 (impact modified and 15%–30% glass fiber)/PA	1.25–1.35
Nylon 6,6 (30% mineral filled)/PA	1.35–1.38
Polystyrene (expanded foam)/EPS	0.01–0.05
Polystyrene (extruded foam)/XPS	0.03–0.05
Polychloroprene (neoprene) (solid)/ CR	1.20–1.24
Polychloroprene (neoprene) (foamed)/CR	0.11–0.56
Polyethylene terephthalate/PET	1.30–1.40
Polyethylene terephthalate (30% glass fiber)/PET	1.50–1.60
Polyethylene terephthalate (30% glass fiber and impact modified)/PET	1.40–1.50
Polycarbonate (high heat)/PC	1.15–1.20
Polycarbonate (20%–40% glass fiber and flame retardant)/PC	1.40–1.50
Polycarbonate (20%–40% glass fiber)/PC	1.35–1.52
Polystyrene (30% glass fiber)/PS	1.40–1.50
Polyvinyl chloride (rigid)/PVC	1.35–1.50
Polyvinyl chloride (20% glass fiber)/PVC	1.45–1.50
Polyvinyl chloride (plasticized)/PVC	1.30–1.70
Polyvinyl chloride (plasticized and filled)/PVC	1.15–1.35

(Continued)

TABLE 7.2 (*Continued*)

List of Some Common Plastics with Density Ranges and Common High-Density Solutions to Separate Sediment Microplastics

Salt Solution/ Chemical Formula	Density (g cm^{-3})
Sodium iodide/NaI	1.8
Zinc chloride/ZnCl$_2$	1.5–1.7
Sodium polytungstate/Na$_6$(H$_2$W$_{12}$O$_{40}$)	1.4
lithium metatungstate (5.4 M)/Li$_6$(H$_2$W$_{12}$O$_{40}$)	1.6

TABLE 7.3

Chemical Reagents Used for Density Separation Along with Their Cost, Density, Solubility, and Toxicity

Reagent	Density (g cm^{-3})	Water Solubility (g L^{-1})	Toxicity/Cost ($)	References
Sodium chloride (NaCl)	1.0–1.2	358 at 20°C	Low/$34.64 for 1 kg	Nuelle et al. (2014); Frias et al. (2018); Campanale et al. (2019); Corcoran et al. (2009)
Sodium tungstate dehydrate (Na$_2$WO$_4$ 2H$_2$O)	1.40	742 at 25°C	Low/$224.92 for 500 g	Frias et al. (2018);
Sodium bromide (NaBr)	1.37–1.40	905 at 20°C	Low/$118.51 for 1 kg	Prata et al. (2019); Quinn et al. (2017); Frias et al. (2018)
Sodium polytungstate (3Na$_2$WO$_4$·9WO$_3$·H$_2$O)	1.40	3,100 at 20°C	Low/$235.27 for 100 g	Frias et al. (2018); Corcoran et al. (2009)
Zinc chloride (ZnCl$_2$)	1.6–1.8	4,320 at 20°C	High/$143.34 for 1 kg	Stock et al. (2019); Nuelle et al. (2014); Masura et al. (2015); Frias et al. (2018); Coppock et al. (2017)
Zinc bromide (ZnBr$_2$)	1.71	4,470 at 20°C	High/$166 for 500 g	Prata et al. (2019); Quinn et al. (2017); Frias et al. (2018)
Sodium iodide (NaI)	1.80	1,793 at 20°C	High/$159.46 for 500 g	Stock et al. (2019); Frias et al. (2018); Coppock et al. (2017); Kedzierski et al. (2018)

To overcome these issues, some researchers prefer to reduce the sample volume before density separation by elutriation and froth flotation can be used (Olencycz et al. 2015; Cole and Sherrington 2016).

7.2.2 ELUTRIATION AS SEPARATION TECHNIQUE

In this process, the particles' separation is based on their size, shape, and density by using a gaseous or liquid stream that flows in an opposite direction to the sedimentation flow as indicated in Figure 7.4.

In microplastics' separation, this technique was first used to separate from sediment by directing an upward flow of water through a column, thereby inducing fluidization of the sediment. Later on, the process was found to be suitable as a pre-treatment step to reduce sample volume before density separation. It showed an excellent efficiency for PVC particles (100%) and fibers (98%) and allowed for the reduction of sediment mass up to 80%. The combination of the elutriation-based AIO technique and density separation with sodium iodide resulted in the recovery of 91–99% of ~1 mm polypropylene, polyvinyl chloride, polyethylene terephthalate, polystyrene and polyurethane microplastics (Olencycz et al. 2015; Cole and Sherrington 2016).

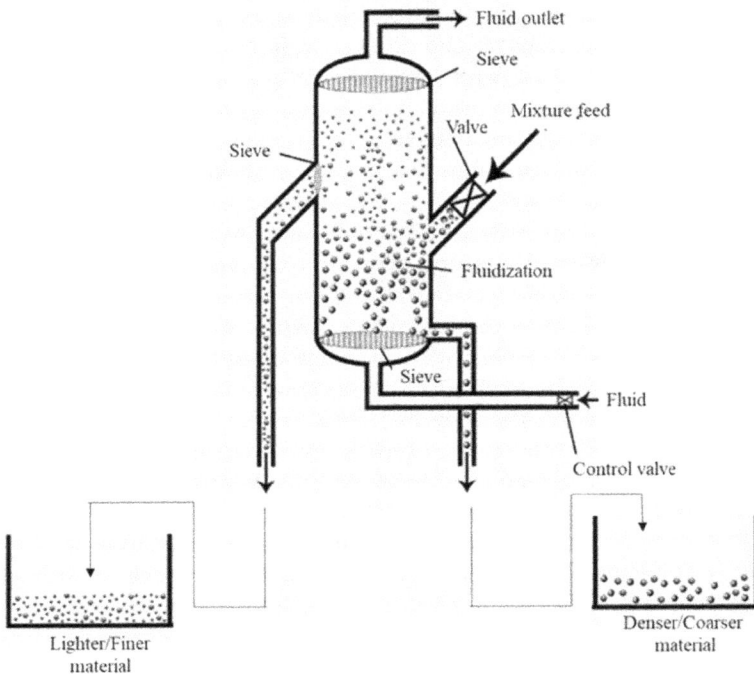

FIGURE 7.4 General sketch of elutriation process for separation of various plastics. (Significantly modified from Crawford and Quinn, 2017.)

7.2.3 FROTH FLOTATION AND OIL FILM SEPARATION TECHNIQUES

In this process, the selective separation of materials is based upon their hydrophobic or hydrophilic characteristics. The basic of froth flotation is surface wettability: materials with a hydrophobic surface tend to float as bubble aggregations, while hydrophilic counterparts are transported as underflow. Post-consumer plastic generally performs a hydrophobic low-energy surface and is entrained with the bubbles group. The hydrophobic particles attached with air bubbles are carried to the surface, thereby forming a froth which can be removed, while hydrophilic materials stay in the liquid phase (Martins and Sobral 2011). Plastics being a hydrophobic material can be separated using the froth flotation technique. Two plastic materials can be separated from one another by adding a wetting agent which can be selectively adsorbed to one of the plastic surfaces. Here, the wetting agent acts as a flotation depressant to promote selective adsorption, thereby rendering the hydrophilic properties. Thus, the hydrophobic plastic will float on upper layers while hydrophilic plastic will sink as a result of depressed flotation, making it possible to extract plastics from the surface.

There are several influential factors that accounted for froth flotation, like the surface free energy of microplastics and the surface tension of the liquid in a flotation bath, as well as the critical surface tension (at which the liquid completely wets the solid microplastics). Further, the selective separation of inherently hydrophobic microplastics requires that the microplastics are only partially wetted by the liquid in the flotation bath, thereby allowing the bubble to adhere to the surface of the solid phase and bring the microplastics to the surface of the liquid phase to be collected, while the sediment particles are completely wetted and sink. Nevertheless, the froth-flotation technique has rarely been employed for microplastics' separation from sediments (Imhof et al. 2013; Possatto et al. 2011) most probably due to the technique has a negative effect by the presence of additives or, wetting agents.

Oil film separation for MPs is a hydrophobicity-based method free from the effect of density. This density-independent approach provided a cost-effective and low-risk alternative to other methods (Figure 7.5). Similarly, Mani et al. used castor oil to separate MPs from the aquatic and sediment samples. The mean spike recovery of MPs was ca. 99%, with an average matrix reduction of 95% (Mani et al. 2019). This advance may lead to a breakthrough in methodical homogeneity and accelerate the removal strategies of MPs. However, Lares et al. (2019) pointed out that solid samples could obstruct separatory funnels easily during separation. The separatory funnel restricted the upper size of recovered MPs when hydrophilic particles were discharged as underflow. Therefore, Wang et al. developed an olive oil-based method to extract MPs from soil and compost samples that consisted of polyethylene, polyurethane, polycarbonate, polyethylene terephthalate, polyvinyl chloride, and polystyrene (Wang et al. 2020). He modified the oil film separation by freezing samples at −40°C in polytetrafluoroethylene cylinders, and the MPs recovery from artificial samples is over 90%. But oil film separation may be hampered by residual oil traces that need an extra clean strategy by ethyl alcohol and hexane (Figure 7.5).

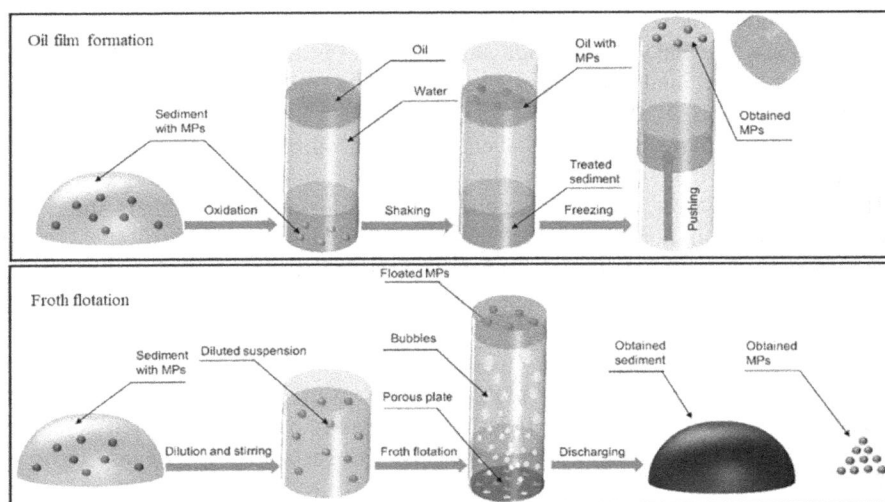

FIGURE 7.5   General sketch of froth flotation and oil film separation techniques for micro-plastics and nanoplastics. (Adopted from Zhang et al., 2021 after permission.)

7.3   SEPARATION OF MICROPLASTICS FROM BIOLOGICAL SAMPLES BY VARIOUS TECHNIQUES AND ASSOCIATED CHALLENGES

The uptake of microplastics and their ingestion by many marine and freshwater microorganisms is widely reported. Indeed, invertebrates (especially bivalves and crustaceans), fish, marine mammals, and birds are known to regularly ingest these materials, and consequently, there has been a general focus on the analysis of gut contents. Only small mini microplastics are observed in organism's tissue which is more likely translocated from the gastrointestinal tract (Giari et al. 2015). Alternative techniques are therefore developed for microplastics' separation from a biological material and are found to be dependent on microorganism's size and the type of microplastics.

7.3.1   VISUAL IDENTIFICATION AS SEPARATION TECHNIQUE

This is a common technique used for detecting and separating the microplastics from biological materials, where it does not require any specialist equipment, other than a binocular microscope. Primarily, it is used to identify and classify the microplastics via their observations on a filter (Song et al. 2015) or in Petri dishes or jars (Reisser et al. 2013) using different microscopes, including optical (Bagaev et al. 2018), fluo-rescence (Cai et al., 2017), electron (Leslie et al. 2011), dissection (Sagawa et al. 2018), stereo (Zobkov et al. 2019), inverted (Gorokhova 2015), binocular (de Lucia et al. 2014), and vertical (Setälä et al. 2016) microscopes. Here, the magnification used for identification varies from ×4.5 to ×400 (Norén 2007; MERI 2017). Due to a higher resolution, the scanning electron microscope can also be used, allowing a precise identification even when distinguishing the color is difficult (Sagawa et al. 2018). To facilitate visual analysis using microscope, the samples are generally placed on gridded filters (MERI 2017).

Three criteria have been established for MP recognition, which allows for differentiating MPs from other materials, in particular, biological matter, as follows: MP particles must not have visible cellular or organic structures; fibers should be equally thick throughout their entire length; and particles should exhibit clear and homogeneous color throughout (Norén 2007; Hidalgo-Ruz et al. 2012; MERI 2017). If these criteria are followed, particles can be defined as plastic (Hidalgo-Ruz et al. 2012).

A melting test and hot needle test can be performed to assess if the observed particles are effectively made of plastic (Enders et al. 2015; Tunçer et al. 2018). These tests consist of burning particles and the burned particle is defined as plastic if it melts or wrinkles (MERI 2017). Because the performed tests cause the loss of analyzed particles, the tests are carried out only for the particles of uncertain nature.

7.3.2 Chemical and Enzymatic Digestion

While visual identification is a mandatory step for the separation of debris and naturally occurring organic fragments, it is impractical to exclusively rely on it. In particular, when dealing with dense material or, minute organisms like zooplanktons or some complex environmental metrics. Therefore, the developed digestion technique allows a faster separation of plastic materials from organic matter (Miller et al. 2017) and is used to degrade the organic portions wherein the microplastics are not removed. The different types of digestion are accomplished with enzymatic (Saliu et al. 2018), acidic and alkaline (Zobkov et al. 2019; Zhu et al. 2019), and oxidative (Pan et al. 2019). Alternately, they are also used in combination (Tamminga et al. 2018) with a common purpose of degrading a wider range of compounds after microplastic separation and coupled with density separation (Masura et al. 2015). This approach is more suitable for a smaller organism where the entire body or visceral mass may be digested to ensure the collection of all plastic materials. Nonetheless, specific cares are to be taken as the digestion can have a chemical impact on the microplastics itself, particularly with fibers.

The commonly used compounds in digestion process are hydrogen peroxide (H_2O_2) alone and/or, combined with a catalyst like $FeSO_4$ (Dai et al. 2018; Masura et al. 2015), hydrofluoric acid (HF) (Dubaish and Liebezeit 2013), hydrochloric acid (HCl) (Zobkov et al. 2019), sodium hypochlorite (NaClO) (Beer et al. 2018), potassium hydroxide (KOH) (Beer et al. 2018) and/or sodium hydroxide (NaOH) (Cole et al. 2014) and different enzymes, of which the proteinase K is a common one (Saliu et al. 2018). Also, they are used in combination, like Dubaish and Liebezeit (2013) used 30% H_2O_2 with 40% HF. After a digestion, the microplastic samples are passed through other separation processes (like filtration). However, the use of strong mineral acids (HCl, HF) is not recommended due to the adverse effect on structural integrity of the microplastics (Miller et al. 2017). In contrast, the acid destruction using a mixture of 65% nitric acid (HNO_3) and 68% perchloric acid ($HClO_4$) at a ratio of 4:1 completely digests the tissues and removes all other organic materials, leaving only plastics and silica (Vesilind 2003). It also removes rayon fibers of regenerated cellulose; however, this technique is still developing, and the effect of acid ratios is interesting to determine as it may have detrimental effects on nylon fibers (Vesilind 2003), which are usually sensitive to acids/alkalis. Table 7.4 indicates the comparison of different reagents used for digestion of organic fraction to separate microplastics and nanoplastics from the complex environment and associated challenges.

7.4 SEPARATION OF MICROPLASTICS FROM ATMOSPHERIC SAMPLES BY VARIOUS TECHNIQUES AND ASSOCIATED CHALLENGES

1. Early atmospheric microplastic research involved a simple visual microscopic reporting of plastic presence and quantification (Dris et al., 2015). While effective for large obvious microplastic particles, it can be difficult to accurately determine if particulates are plastics when considering particles <500 μm (Hidalgo-Ruz et al. 2012).
2. The valid sample preparation methods are yet to be standardized. For which, numerous organic removals including density separation methods are used. Here, the density separation requires material to be suspended or settled in liquid of various densities. It has been carried out with freshwater (1.0 g mL^{-1}), seawater (1.03 g mL^{-1}), sodium chloride (>1.2 g mL^{-1}), calcium chloride (>1.35 g mL^{-1}), sodium polytungstate solution (>1.5 g mL^{-1}), sodium bromide (>1.6 g mL^{-1}), zinc bromide (>1.7 g mL^{-1}), zinc chloride (>1.7 g mL^{-1}), and sodium iodide (>1.8 g mL^{-1}) (Li et al. 2018; Crawford and Quinn 2017). Notably, to settle fine dust particles in atmospheric deposition (e.g., Saharan dust and similar material), it has been required to lightly agitate the settling tubes (60 rpm) for preventing a collation of fine dusts on and around microplastic particles.
3. The organic removal through different digestive methods can be done including KOH, NaOH, HNO_3, HCl, H_2O_2, $H_2O_2 + H_2SO_4$, $H_2O_2 + Fe$, and enzymatic methods. Some atmospheric deposition studies have also used sodium hypochlorite (NaClO) or hydrogen peroxide (H_2O_2) as digestion methods for organic removal to date (Allen et al. 2019; Klein and Fischer 2019; Hanvey et al. 2017; Löder et al. 2015). In recent times, Fenton's reagent has been identified as an effective advancement to sample preparation processes (Hurley et al. 2018; Prata et al. 2019).

ACKNOWLEDGMENTS

This work was supported by a grant from the National Research Foundation of Korea (NRF) grant funded by the Korea government (MSIT) (No. NRF-2020R1A2C1013851) and the National Research Foundation of Korea (NRF) funded by the Ministry of Education (2021R1I1A1A01054655).

TABLE 7.4

Comparison of Different Reagents Used for Digestion of Organic Fraction to Separate Microplastics and Nanoplastics from Complex Environment

Reagents	Experimental Conditions	Merits	Costs	Associated Challenge	References
HNO_3	20 mL of HNO_3 (22.5 M), 2 h heating (~100°C), hot filtration (~80°C)	Efficient in organic digestion	$38.00 for 1 L	Oxidizer, Corrosive, degradation of PS, PA, and PE, makes the plastic yellow	Stock et al. (2019); Claessens et al. (2013); Avio et al. (2015); Dehaut et al. (2016)
HCl	4 mL of HCl at 20%	Efficient in organic digestion (82.6%) of complex matrices (clams)	$36.00 for 1 L	Corrosive, acute toxicity, Degradation of polymers	Cole et al. (2014); Thiele et al. (2019); Nuelle et al. (2014)
NaOH	20 mL of NaOH (10 M) at 60°C for 24 h	Digestion efficiency up to 90%, stimulated by the rise of molarity and temperature	$62.63 for 1 kg	Corrosive, Degradation of PET and PVC	Stock et al. (2019); Cole et al. (2014); Nuelle et al. (2014); Hurley et al. (2018)
KOH	20 mL of KOH (1 M) at 18°C–21°C for two days	Good organic digestion efficiency	$85.72 for 1 kg	Corrosive, Time consuming, degradation of cellulose acetate and some biodegradable plastics	Stock et al. (2019); Cole et al. (2014); Dehaut et al. (2016); Kühn et al. (2017)
H_2O_2	20 mL H_2O_2 at 30% plus 20 mL of $FeSO_4 \cdot 7H_2O$ (0.05M) at 70°C in stirring	Efficient in organic digestion	$27.67 for 1 L	Corrosive, At high concentrations could degrade the polymers	Campanale et al. (2020); Stock et al. (2019); Nuelle et al. (2014); Masura et al. (2015)

(Continued)

TABLE 7.4 (Continued)

Comparison of Different Reagents Used for Digestion of Organic Fraction to Separate Microplastics and Nanoplastics from Complex Environment

Reagents	Experimental Conditions	Merits	Costs	Associated Challenge	References
Cellulase, lipase, chitinase, protease, proteinase-K	5 mL of Protease A-01 + 25 mL of Tris-HCl buffer; 1 mL of Lipase FE-01 + 25 mL of Tris-HCl buffer; 5 mL of Amylase TXL + 25 mL of NaOAc buffer, 1 mL of Cellulase TXL + 25 mL of NaOAc buffer; 1 mL of Chitinase + 25 mL of NaOAc buffer	Good inorganic and biological material digestion; does not affect the polymers	Protease A-01 1 kg $48.34; Lipase FE-01 1 kg $48.34; Amylase TXL 1 kg $36.50; Cellulase TXL 1 kg $43.95	Expensive, time consuming	Li et al. (2018); Stock et al. (2019); Cole et al. (2014); Löder et al. (2017)

REFERENCES

Abayomi, O. A., Range, P., Al-Ghouti, M. A., Obbard, J. P., Almeer, S. H., Ben-Hamadou, R. (2017) Microplastics in coastal environments of the Arabian Gulf. *Marine Pollution Bulletin*, 124, 181–188.

Avio, C. G., Gorbi, S., Regoli, F. (2015) Experimental development of a new protocol for extraction and characterization of microplastics in fish tissues: first observations in commercial species from Adriatic Sea. *Marine Environmental Research*, 111, 18–26.

Allen, S., Allen, D., Phoenix, V. R., Le Roux, G., Jiménez, P. D., Simonneau, A., Galop, D. (2019) Atmospheric transport and deposition of microplastics in a remote mountain catchment. *Nature Geoscience*, 12(5), 339–344.

Bagaev, A., Khatmullina, L., Chubarenko, I. (2018) Anthropogenic microlitter in the Baltic Sea water column. *Marine Pollution Bulletin*, 129(2), 918–923.

Barrows, A. P. W., Neumann, C. A., Berger, M. L., Shaw, S. D. (2017) Grab *vs.* neuston tow net: a microplastic sampling performance comparison and possible advances in the field. *Anal Methods-UK*, 9, 1446.

Beer, S., Garm, A., Huwer, B., Dierking, J., Nielsen, T. G. (2018) No increase in marine microplastic concentration over the last three decades – a case study from the Baltic Sea. *Science of the Total Environment*, 621, 1272–1279.

Castillo, A. B., Al-Maslamani, I., Obbard, J. P. (2016) Prevalence of microplastics in the marine waters of Qatar. *Marine Pollution Bulletin*, 111(1–2), 260–267.

Campanale, C., Massarelli, C., Bagnuolo, G., Savino, I., Uricchio, V. F. (2019) The problem of microplastics and regulatory strategies in Italy. In; Stock, F., Reifferscheid, G., Brennholt, N., Kostianaia, E., (Eds.). *Plastics in the Aquatic Environment-Stakeholders Role against Pollution.* Springer: Cham, Switzerland.

Campanale, C., Stock, F., Massarelli, C., Kochleus, C., Bagnuolo, G., Reifferscheid, G., Uricchio, V. F. (2020) Microplastics and their possible sources: The example of Ofanto river in southeast Italy. *Environmental Pollution*, 258, 113284.

Castro, R. O., Silva, M. L., Marques, M. R. C., de Araújo, F. V. (2016) Evaluation of microplastics in Jurujuba Cove, Niterói, RJ, Brazil, an area of mussels farming. *Marine Pollution Bulletin*, 110, 555–558.

Cai, L., Wang, J., Peng, J., Tan, Z., Zhan, Z., Tan, X. & Chen, Q. (2017) Characteristic of microplastics in the atmospheric fallout from Dongguan city, China: preliminary research and first evidence. *Environmental Science and Pollution Research*, 24(32), 24928–24935.

Claessens, M., Van Cauwenberghe, L., Vandegehuchte, M. B., Janssen, C. R. (2013) New techniques for the detection of microplastics in sediments and field collected organisms. *Marine Pollution Bulletin*, 70(1–2), 227–233.

Cole, M., Webb, H., Lindeque, P. K., Fileman, E. S., Halsband, C., Galloway, T. S. (2014) Isolation of microplastics in biota-rich seawater samples and marine organisms. *Scientific Reports*, 4(1), 1–8.

Cole, G., Sherrington, C. (2016) *Study to Quantify Pellet Emissions in the UK.* Eunomia: Bristol.

Corcoran, P. L., Norris, T., Ceccanese, T., Walzak, M. J., Helm, P. A., Marvin, C. H. (2015) Hidden plastics of Lake Ontario, Canada and their potential preservation in the sediment record. *Environmental Pollution*, 204, 17–25.

Crawford, C. B., Quinn, B. (2017) *Microplastic Pollutants*, Elsevier: Amsterdam, The Netherland. ISBN: 9780128094068.

Coppock, R. L., Cole, M., Lindeque, P. K., Queirós, A. M., Galloway, T. S. (2017) A small-scale, portable method for extracting microplastics from marine sediments. *Environmental Pollution*, 230, 829–837.

Corcoran, P. L., Biesinger, M. C., Grifi, M. (2009) Plastics and beaches: a degrading relationship. *Marine Pollution Bulletin*, 58(1), 80–84.

Cutroneo, L., Reboa, A., Besio, G., Borgogno, F., Canesi, L., Canuto, S., Capello, M. (2020) Microplastics in seawater: sampling strategies, laboratory methodologies, and identification techniques applied to port environment. *Environmental Science and Pollution Research*, 27(9), 8938–8952.

Dai, Z., Zhang, H., Zhou, Q., Tian, Y., Chen, T., Tu, C., Fu, C., Luo, Y. (2018) Occurrence of microplastics in the water column and sediment in an inland sea affected by intensive anthropogenic activities. *Environmental Pollution*, 242, 1557–1565.

Desforges, J. P. W., Galbraith, M., Dangerfield, N., Ross, P. S. (2014) Widespread distribution of microplastics in subsurface seawater in the NE Pacific Ocean. *Marine Pollution Bulletin*, 79(1–2), 94–99.

Dehaut, A., Cassone, A. L., Frère, L., Hermabessiere, L., Himber, C., Rinnert, E., Paul-Pont, I. (2016) Microplastics in seafood: Benchmark protocol for their extraction and characterization. *Environmental Pollution*, 215, 223–233.

de Lucia, G. A., Caliani, I., Marra, S., Camedda, A., Coppa, S., Alcaro, L., Campani, T., Giannetti, M., Coppola, D., Cicero, A, M., Panti, C., Baini, M., Guerranti, C., Marsili, L., Massaro, G., Fossi, M. C., Matiddi, M. (2014) Amount and distribution of neustonic micro-plastic off the western Sardinian coast (Central-Western Mediterranean Sea). *Marine Environmental Research*, 100, 10–16.

Dris, R., Gasperi, J., Rocher, V., Saad, M., Renault, N., Tassin, B. (2015) Microplastic contamination in an urban area: A case study in Greater Paris. *Environmental Chemistry*, 12, 592–599.

Dubaish, F., Liebezeit, G. (2013) Suspended microplastics and black carbon particles in the jade system, southern North Sea. *Water, Air, & Soil Pollution*, 224(2), 1–8.

Enders, K., Lenz, R., Stedmon, C. A., Nielsen, T. G. (2015) Abundance, size and polymer composition of marine microplastics≥ 10 μm in the Atlantic Ocean and their modelled vertical distribution. *Marine Pollution Bulletin*, 100(1), 70–81.

Erni-Cassola, G., Gibson, M, I., Thompson, R, C., Christie-Oleza, J. A. (2017) Lost, but found with Nile red: a novel method for detecting and quantifying small microplastics (1 mm to 20 μm) in environmental samples. *Environmental Science & Technology*, 51, 13641–13648.

Eriksen, M., Mason, S., Wilson, S., Box, C., Amato, S. (2013) Microplastic pollution in the surface waters of the Laurentian Great Lakes. *Marine Pollution Bulletin*, 77, 177–182. doi: 10.1016/j.marpolbul.2013.10.007.

Figueiredo, G. M., Vianna, T. M. P. (2018) Suspended microplastics in a highly polluted bay: Abundance, size, and availability for mesozooplankton. *Marine Pollution Bulletin*, 135, 256–265.

Frias, J. P., Nash, R., Pagter, E., O'Connor, I. (2018) Standardised Protocol for Monitoring Microplastics in Sediments; Technical Report; BASEMAN Project; JPI-Oceans: Brussels, Belgium.

Giari, L., Guerranti, C., Perra, G., Lanzoni, M., Fano, E. A., Castaldelli, G. (2015) Occurrence of perfluorooctanesulfonate and perfluorooctanoic acid and histopathology in eels from north Italian waters. *Chemosphere*, 118, 117–123.

Gorokhova, E. (2015) Screening for microplastic particles in plankton samples: how to integrate marine litter assessment into existing monitoring programs? *Marine Pollution Bulletin*, 99(1-2), 271–275.

Hall, N. M., Berry, K. L. E., Rintoul, L., Hoogenboom, M. O. (2015) Microplastic ingestion by scleractinian corals. *Marine Biology*, 162(3), 725–732.

Hanvey, J. S., Lewis, P. J., Lavers, J. L., Crosbie, N. D., Pozo, K., Clarke, B. O. (2017) A review of analytical techniques for quantifying microplastics in sediments. *Analytical Methods*, 9(9), 1369–1383.

Hidalgo-Ruz, V., Gutow, L., Thompson, R. C., Thiel, M. (2012) Microplastics in the marine environment: a review of the methods used for identification and quantification. *Environmental Science & Technology*, 46(6), 3060–3075.

Hidalgo-Ruz, V., Thiel, M. (2013) Distribution and abundance of small plastic debris on beaches in the SE Pacific (Chile): A study supported by a citizen science project. *Marine Environmental Research*, 87–88, 12–18.

Hurley, R., Woodward, J., Rothwell, J. J. (2018) Microplastic contamination of river beds significantly reduced by catchment-wide flooding. *Nature Geoscience*, 11(4), 251–257.

Imhof, H. K., Ivleva, N. P., Schmid, J., Niessner, R., Laforsch, C. (2013) Contamination of beach sediments of a subalpine lake with microplastic particles. *Current Biology*, 23(19), R867–R868.

Kang, J. H., Kwon, O. Y., Lee, K. W., Song, Y. K., Shim, W. J. (2015) Marine neustonic microplastics around the southeastern coast of Korea. *Marine Pollution Bulletin*, 96(1–2), 304–312.

Kedzierski, M.,Tilly, V.L., Bourseau, P., Bellegou, H., César, G., Sire, O., Bruzaud, S. 2017. Microplastic elutriation system. Part A: Numerical modeling. Marine Pollution Bulletin. 119, 151–161.

Klein, M., Fischer, E. K. (2019) Microplastic abundance in atmospheric deposition within the Metropolitan area of Hamburg, Germany. *Science of the Total Environment*, 685, 96–103.

Kroon, F., Motti, C., Talbot, S., Sobral, P., Puotinen, M. (2018) A workflow for improving estimates of microplastic contamination in marine waters: a case study from North-Western Australia. *Environmental Pollution*, 238, 26–38.

Kühn, S., Van Werven, B., Van Oyen, A., Meijboom, A., Rebolledo, E. L. B., Van Franeker, J. A. (2017) The use of potassium hydroxide (KOH) solution as a suitable approach to isolate plastics ingested by marine organisms. *Marine Pollution Bulletin*, 115(1–2), 86–90.

Leslie, H. A., van der Meulen, M. D., Kleissen, F. M., Vethaak, A. D. (2011) Microplastic litter in the Dutch marine environment: providing facts and analysis for Dutch policymakers concerned with marine microplastic litter. (Rapport 1203772-000). Deltares / IVM-VU: Delft / Amsterdam.

Löder, M. G. J., Kuczera, M., Mintenig, S., Lorenz, C., Gerdts, G. (2015) Focal plane array detector-based micro-Fourier-transform infrared imaging for the analysis of microplastics in environmental samples. *Environmental Chemistry*, 12(5), 563–581.

Lusher, A. L., Welden, N. A., Sobral, P., Cole, M. (2017) Sampling, isolating and identifying microplastics ingested by fish and invertebrates. *Analytical Methods*, 9, 1346–1360.

Li, J., Liu, H., Chen, J. P. (2018) Microplastics in freshwater systems: A review on occurrence, environmental effects, and methods for microplastics detection. *Water Research*, 137, 362–374.

Lima, A. R. A., Costa, M. F., Barletta, M. (2014) Distribution patterns of microplastics within the plankton of a tropical estuary. *Environmental Research*, 132, 146–155.

Lares, M., Ncibi, M.C., Sillanpää, M. (2019) Intercomparison study on commonly used methods to determine microplastics in wastewater and sludge samples. *Environmental Science and Pollution Research*, 26, 12109–12122.

Mani, T., Frehland, S., Kalberer, A., Burkhardt-Holm, P. (2019) Using castor oil to separate microplastics from four different environmental matrices. *Analytical Methods*, 11, 1788–1794.

Maes, T., Jessop, R., Wellner, N., Haupt, K., Mayes, A. G. (2017) A rapid-screening approach to detect and quantify microplastics based on fluorescent tagging with Nile Red. *Scientific Reports*, 7(1), 1–10.

Martins, J., Sobral, P. (2011) Plastic marine debris on the Portuguese coastline: A matter of size? *Marine Pollution Bulletin*, 62(12), 2649–2653.

Masura, J., Baker, J., Foster, G., Arthur, C. (2015) Laboratory Methods for the Analysis of Microplastics in the Marine Environment: Recommendations for quantifying synthetic particles in waters and sediments. Available from: https://marinedebris.noaa.gov/sites/default/files/publications-files/noaa_microplastics_methods_manual.pdf.

MERI (Marine & Environmental Research Institute). (2017) Guide to microplastic identification. Marine & Environmental Research Institute, p 14. https://www.ccb.se/documents/Postkod2017/Mtg050317/Guide%20to%20Microplastic%20Identification_MERI.pdf.

Miller, M. E., Kroon, F. J., Motti, C. A. (2017) Recovering microplastics from marine samples: a review of current practices. *Marine Pollution Bulletin*, 123(1–2), 6–18.

Norén, F. (2007) Small plastic particles in coastal Swedish waters. *Kimo Sweden*, 11, 1–11.

Nuelle, M. T., Dekiff, J. H., Remy, D., Fries, E. (2014) A new analytical approach for monitoring microplastics in marine sediments. *Environmental Pollution*, 184, 161–169.

Olencycz, M., Sokołowski, A., Niewińska, A., Wołowicz, M., Namieśnik, J., Hummel, H., Jansen, J. (2015) Comparison of PCBs and PAHs levels in European coastal waters using mussels from the Mytilus edulis complex as biomonitors. *Oceanologia*, 57(2), 196–211.

Pan, Z., Guo, H., Chen, H., Wang, S., Sun, X., Zou, Q., Huang, J. (2019) Microplastics in the Northwestern Pacific: Abundance, distribution, and characteristics. *Science of the Total Environment*, 650, 1913–1922.

Prata, J. C., da Costa, J. P., Duarte, A. C., Rocha-Santos, T. (2019) Methods for sampling and detection of microplastics in water and sediment: A critical review. *TrAC Trends in Analytical Chemistry*, 110, 150–159.

Possatto, F. E., Barletta, M., Costa, M. F., Ivar do Sul, J. A., Dantas, D. V. (2011) Plastic debris ingestion by marine catfish: An unexpected fisheries impact. *Marine Pollution Bulletin*, 62(5), 1098–1102.

Quinn, B., Murphy, F., Ewins, C. (2017) Validation of density separation for the rapid recovery of microplastics from sediment. *Analytical Methods*, 9, 1491–1498. doi: 10.1039/c6ay02542k.

Reisser, J., Shaw, J., Wilcox, C., Hardesty, B. D., Proietti, M., Thums, M., Pattiaratchi, C. (2013) Marine plastic pollution in waters around Australia: characteristics, concentrations, and pathways. *PLoS One*, 8(11), e80466.

Sagawa, N., Kawaai, K., Hinata, H. (2018) Abundance and size of microplastics in a coastal sea: comparison among bottom sediment, beach sediment, and surface water. *Marine Pollution Bulletin*, 133, 532–542.

Saliu, F., Montano, S., Garavaglia, M, G., Lasagni, M., Seveso, D., Galli, P. (2018) Microplastic and charred microplastic in the Faafu Atoll, Maldives. *Marine Pollution Bulletin*, 136, 464–471.

Sigma Aldrich. (2019) Thermal transitions of homopolymers: glass transition & melting point. https://www.sigmaaldrich.com/technical-documents/articles/materials-science/-polymer-science/thermal-transitions-of-homopolymers.html.

Setälä, O., Magnusson, K., Lehtiniemi, M., Norén, F. (2016) Distribution and abundance of surface water microlitter in the Baltic Sea: a comparison of two sampling methods. *Marine Pollution Bulletin*, 110(1), 177–183.

Song, Y. K., Hong, S. H., Jang, M., Han, G. M., Rani, M., Lee, J., Shim, W. J. (2015) A comparison of microscopic and spectroscopic identification methods for analysis of microplastics in environmental samples. *Marine Pollution Bulletin*, 93(1–2), 202–209.

Suaria, G., Avio, C. G., Mineo, A., Lattin, G. L., Magaldi, M. G., Belmonte, G. (2016) The Mediterranean plastic soup: synthetic polymers in Mediterranean surface waters. *Scientific Reports-UK*, 6, 37551.

Syakti, A. D., Hidayati, N. V., Jaya, Y. V., Siregar, S. H., Yude, R., Suhendy Asia, L., Wong-Wah-Chung, P., Doumenq, P. (2018) Simultaneous grading of microplastic size sampling in the Small Island of Bintan water, Indonesia. *Marine Pollution Bulletin*, 137, 593–600.

Stock, F., Kochleus, C., Bänsch-Baltruschat, B., Brennholt, N., Reifferscheid, G. (2019) Sampling techniques and preparation methods for microplastic analyses in the aquatic environment–A review. *TrAC Trends in Analytical Chemistry*, 113, 84–92.

Tamminga, M., Hengstmann, E., Fischer, E. K. (2018) Microplastic analysis in the south Funen archipelago, Baltic Sea, implementing manta trawling and bulk sampling. *Marine Pollution Bulletin*, 128, 601–608.

Tang, G., Liu, M., Zhou, Q., He, H., Chen, K., Zhang, H., Hu, J., Huang, Q., Luo, Y., Ke, H., Chen, B., Xu, X., Cai, M. (2018) Microplastics and polycyclic aromatic hydrocarbons (PAHs) in Xiamen coastal areas: implications for anthropogenic impacts. *Science of the Total Environment*, 634, 811–820.

Thiele, C. J., Hudson, M. D., Russell, A. E. (2019) Evaluation of existing methods to extract microplastics from bivalve tissue: Adapted KOH digestion protocol improves filtration at single-digit pore size. *Marine Pollution Bulletin*, 142, 384–393.

Titow, W. V. (1986) *PVC Technology*. Elsevier Applied Science Publishers Ltd. ISBN 978-94-010-8976-0.

Tunçer, S., Artüz, O, B., Demirkol, M., Artüz, M. L. (2018) First report of occurrence, distribution, and composition of microplastics in surface waters of the Sea of Marmara, Turkey. *Marine Pollution Bulletin*, 135, 283–289.

Van der Hal, N., Ariel, A., Angel, D. L. (2017) Exceptionally high abundances of microplastics in the oligotrophic Israeli Mediterranean coastal waters. *Marine Pollution Bulletin*, 116(1–2), 151–155.

Vesilind, P. (2003) *Wastewater Treatment Plant Design*, Vol. 2. IWA Publishing. ISBN 1-84339-029-8.

Viršek, M. K., Palatinus, A., Koren, Š., Peterlin, M., Horvat, P., Kržan, A. (2016) Protocol for microplastics sampling on the sea surface and sample analysis. *Journal of Visualized Experiments*, 118, e55161.

Wang, C., Chelazzi, D., Mikola, J., Leiniö, V., Heikkinen, R., Cincinelli, A. Pellinen, J. (2020) Olive oil-based method for the extraction, quantification and identification of microplastics in soil and compost samples. *Science of the Total Environment*, 733, 139338.

Zhao, S., Zhu, L., Wang, T., Li, D. (2014) Suspended microplastics in the surface water of the Yangtze Estuary System, China: First observations on occurrence, distribution. *Marine Pollution Bulletin*, 86(1–2), 562–568.

Zhang, Y., Jiang, H., Bian, K., Wang, H., Wang, C. (2021) A critical review of control and removal strategies for microplastics from aquatic environments. *Journal of Environmental Chemical Engineering*, 9(4), 105463. ISSN 2213-3437.

Zhu, J., Zhang, Q., Li, Y., Tan, S., Kang, Z., Yu, X., Lan, W., Cai, L., Wang, J., Shi, H. (2019) Microplastic pollution in the Maowei Sea, a typical mariculture bay of China. *Science of the Total Environment*, 658, 62–68.

Zobkov, M. B., Esiukova, E. E., Zyubin, A. Y., Samusev, I. G. (2019) Microplastic content variation in water column: the observations employing a novel sampling tool in stratified Baltic Sea. *Marine Pollution Bulletin*, 138, 193–205.

8 Identification of Microplastics and Nanoplastics and Associated Analytical Challenges

Sadia Ilyas and Hyunjung Kim
Hanyang University

CONTENTS

After doing the sample collection and extraction of microplastics, identifying suspected items that can be composed of plastic particles is an important task. Nevertheless, a precise composition of suspected particles is very broad to be microplastics or nanoplastics. Due to fillers and additives blending at virgin plastic manufacturing or existing as copolymers, the ranges and types of microplastics obtained from a sample (atmospheric/marine environment) vary extensively. Thus, each type of plastic, owing to its specific physicochemical properties, challenges identification accuracy.

Comparisons of protocols for analysis and certified reference materials for various instrumental identification are lacking. There is little information on the recovery

DOI: 10.1201/9781003200628-8

rates of different sampling and processing methods. There is also a profound lack of knowledge regarding the modifications that plastics, and, more precisely, microplastics and nanoplastics undergo once subject to the elements. A further important issue is the contamination of microplastics samples by particles originating, e.g., from the clothing of workers, the used equipment and the ambient air. As far as possible, such contamination has to be reduced. Ultimately, these considerations may lead to the implementation of standardized methodologies for the quantification of microplastics and nanoplastics in the environment (Silva et al., 2018).

8.1   SCANNING ELECTRON MICROSCOPE AS AN IDENTIFICATION TOOL AND ASSOCIATED CHALLENGES

A scanning electron microscope (SEM) creates a clear image of the small surface exposed by firing a high-intensity electron beam at the sample surface and through a raster scanning in a zig-zag pattern. As the electrons are employed in surface imaging, a resolution as low as 0.5 nm is achievable at large magnifications as high as ×2,000,000 times. This magnification is much higher than a standard optical microscope that can go a magnification of ×1,000 times only. In SEM, electrons' generation is achieved either through a hot filament source or through field emission. In the first one, a high electrical current is passed via tungsten filament to generate heat ~5,000°C, which provides sufficient energy to overcome the potential energy barrier. Thus emitting electrons by the thermionic emission from tungsten and characterized by a Maxwell–Boltzmann energy distribution. To ensure electrons reach the sample uninterrupted by the air, the heating is performed in a vacuum. In the second method of field emission, the quantum tunneling of electrons by the potential energy barrier through utilizing a tungsten crystal in the production of electrons possesses a narrow range of energy which supports attaining a high resolution. The incident beam tightly focused by electromagnetic lenses to the sample allows the scanning coils to move the beam over the surfaces of the sample in a raster pattern to scatter the incident (backscattered) electrons. Then the electrons are collected by a detector and transformed into a signal, wherein, a high atomic number element tends to generate larger backscattered electrons. When examining an SEM image, the majority of electrons can break molecular bonds that emit secondary electrons from the sample and tend to a high degree of pertaining to the sample surface. On the other side, X-rays are also generated by hitting the electron beam on to sample surface which can be transformed into voltage with respect to the emission intensity by an X-ray microanalyzer via energy dispersive X-ray analysis, providing a quantitative information on elemental composition (Eriksen et al., 2013; Vandermeersch et al., 2015) (Figure 8.1).

Due to a high depth of field, samples can be visualized in black and white for the morphological, topographical, and compositional information. For a clear visual, the sample surface should be electrically grounded to avoid electrostatic charge building-up. It requires a conductive film coating of low-vacuum sputter-coating or high-vacuum evaporation (typically a thin layer of noble metals, chromium, or graphite), making it an expensive technique. SEM is used for physical analysis of microplastic samples and to measure the size and dimensions of any surface features, to distinguish a polymer item from a non-plastic item. Unlike IR, SEM is not

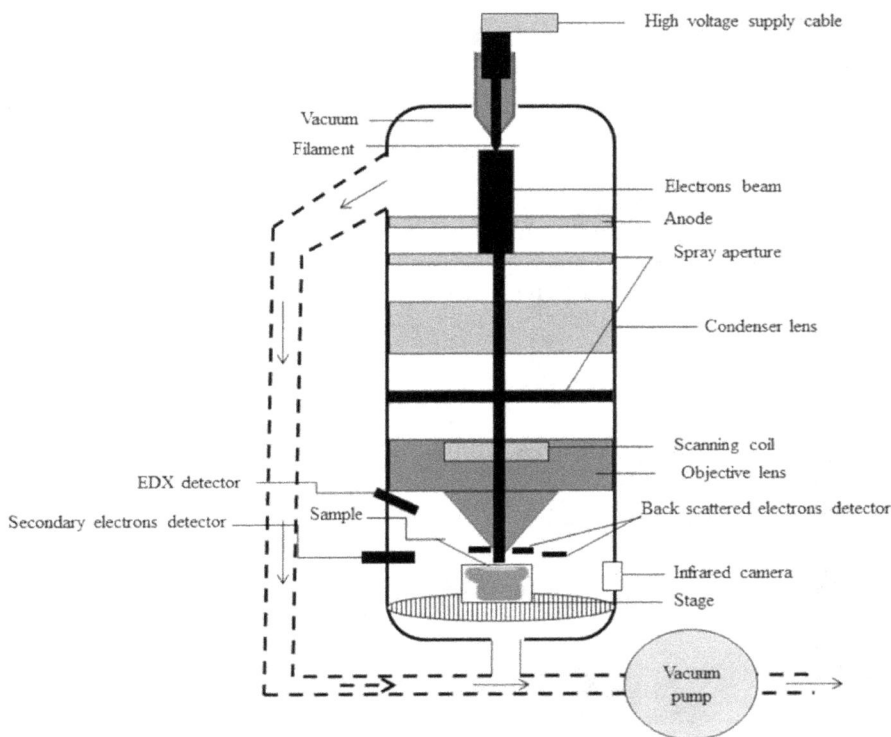

FIGURE 8.1   General sketch of scanning electron microscope.

normally used to identify the type of plastic, albeit SEM equipped with an energy dispersive X-ray can provide the inorganic chemical composition of microplastics (Eriksen et al., 2013; Vandermeersch et al., 2015). Thus, the SEM technique is preferably used to differentiate mini-microplastics (<1 mm–1 µm), including multi-colored microbeads from other materials (Figure 8.1).

In a work conducted by Dehghani et al. (2017) using SEM, the accurate detection of microplastic particles of different shapes (e.g., a spherical, hexagonal, irregular polyhedron, etc.) and size has been performed. At the same time, energy dispersive spectroscopy (EDS) could easily identify Al, Na, Ca, Mg, and Si. In which the chemical composition revealed the presence of polymer additives and/or adsorbed debris onto the surface of the microplastics (Dehghani et al., 2017).

However, no differential signatures during elemental analysis between the polymer additives and the debris adsorbed on microplastics' surface showed the limitation of EDS spectral analysis. The detection of microplastic in wild and farmed mussels was performed by Li et al. (2016). The visual identification of microplastics under a stereo microscope based on their physical characteristics, indicated smooth irregular surface topographies as depicted in Figure 8.2a and b. About 8.5% of the suspected particles, selected by visual identification, were detected by micro-FTIR as diethanolamine and selenious acid; subsequently, some particles of microplastics

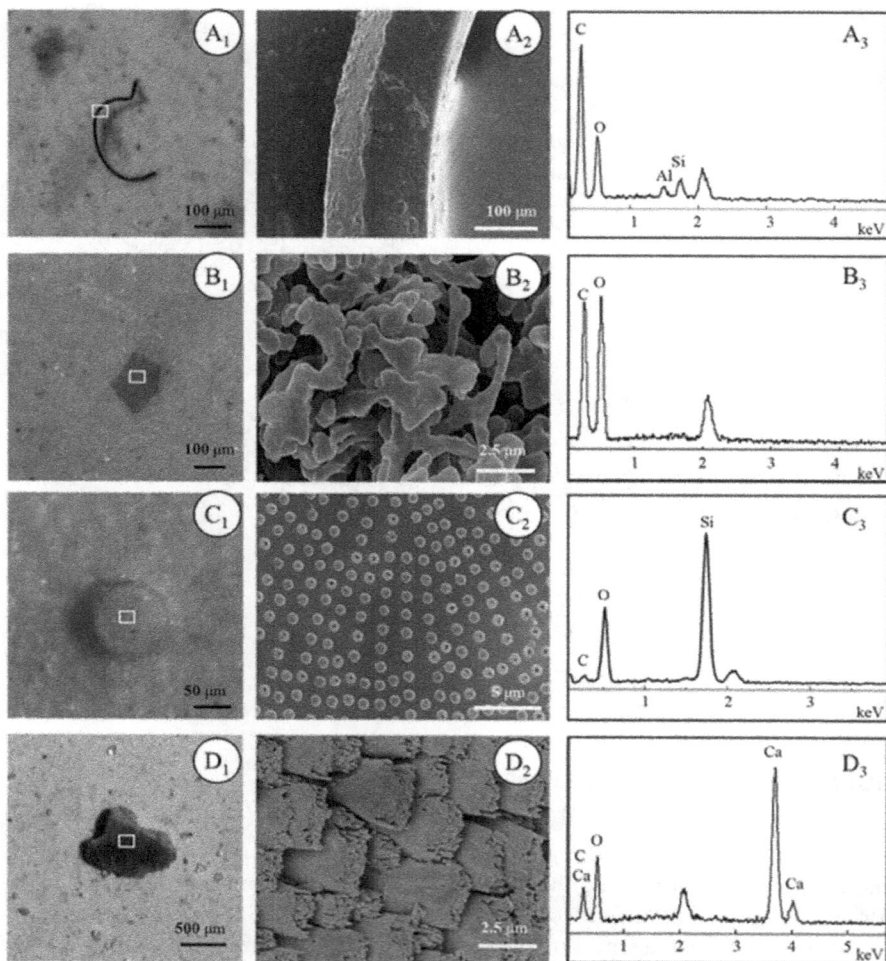

FIGURE 8.2 Identification of microplastics with SEM–EDS. Some particles were identified as microplastics (a and b), and the others were identified as non-plastics such as diatoms (c), and CaCO$_3$ (d). (A1, B1, C1, D1; microscopic view, A2, B2, C2, D2; SEM, A3, B3, C3, D3; EDS). (Adopted from Li et al., 2016 after permission.)

were revealed by SEM–EDS as diatoms (uniform, transparent spheres; Figure 8.2c) and CaCO$_3$ (Figure 8.2d) and about 7% remained unidentified.

Recently, optical microscopy in conjunction with the SEM–EDS has been employed for microplastics the analysis retrieved from ocean trawls and fish guts to determine the surface morphology, size, and chemical composition (Wang et al., 2017). The optical images of microplastics (particles size 70–600 μm) resulted in easy identification of chlorinated plastics (e.g., PVC) due to the unique elemental signature that includes the presence of chlorine, in comparison to the mineral species (Wang et al., 2017). On the other hand, the morphology acquired by optical microscopy and SEM indicated that the fish-ingested particles consist of degraded plastic pieces and microplastics (Wang et al., 2017). Napper and Thompson (Napper and Thompson, 2016) performed a concurrent use of the spectral techniques and observed that SEM–EDS provided complete information than that of using the Raman and FTIR spectroscopies. Micro-FTIR followed by SEM techniques could confirm the release of synthetic fibers like nylon. The release of a greater number of microscopic fibers (<5 mm) of acrylic nature with low quantity for PE-cotton-based fibers was also observed.

Li et al. (2016) found micro-FTIR misleading as it identified aluminum silicate, which was actually revealed to be a diatom by SEM analysis, emphasizing the need for complementary methods to accurately determine the suspected particles (Li et al., 2016). Specifically, SEM can be used to determine the surface morphology like cracks and pits to reveal the degradation pathway of microplastics in the environment (Ter Halle et al., 2017) (Figure 8.3). The surface cracking can also lead to embrittlement, as depicted in Figure 8.3.

8.2 PYROLYSIS–GAS CHROMATOGRAPHY–MASS SPECTROMETRY AS IDENTIFICATION TOOL AND ASSOCIATED CHALLENGES

General schematics of pyrolysis–gas chromatography–mass spectrometry (Py–GC–MS) is presented in Figure 8.4. Py–GC–MS is a destructive technique used for microplastic characterization for identifying the polymer type through the analysis of thermal degradation products (Qiu et al., 2016). It involves an interesting symbiosis that provides a complex mixture of volatile fragments (pyrogram) of the original non-GC amenable sample (or macromolecule) that may be a very effective fingerprint (Picó and Barceló, 2020). Nowadays, different working modes of the Py–GC–MS have been described depending on the purpose of the analysis (Picó and Barceló, 2020);

Single-shot analysis (Py–GC–MS), in which pyrolysis is performed at a single temperature, normally >500°C (dependent on the material being examined). The sample temperature goes as rapid as possible from ambient to the pyrolysis temperature (in current instruments <20 ms). The macromolecules are instantly fragmented in the pyrolyzer, and then, their pyrolyzate are separated in the chromatographic column.

Double-Shot analysis (TD/Py–GC–MS) provides information about both types of compounds (volatile and non-volatile). Volatile (low molecular weight compounds analyzed at low temperature through a thermal desorption step) and non-volatile (investigated at high pyrolysis temperatures that allow for the fragmentation of the macromolecules). Therefore, the analysis of the sample involves two stages: in the first one, there is the thermal desorption of the volatile compounds then, analyzed by GC–MS, and in the second one, the residual sample left after desorption (in which the non-volatile macromolecules remain) is pyrolyzed, and the pyrolyzate is also analyzed by GC–MS.

Evolved gas analysis (EGA–MS) involves separating degradation products from macromolecules according to the temperature at which they are formed rather than their volatilization temperature. This is achieved through a sequential

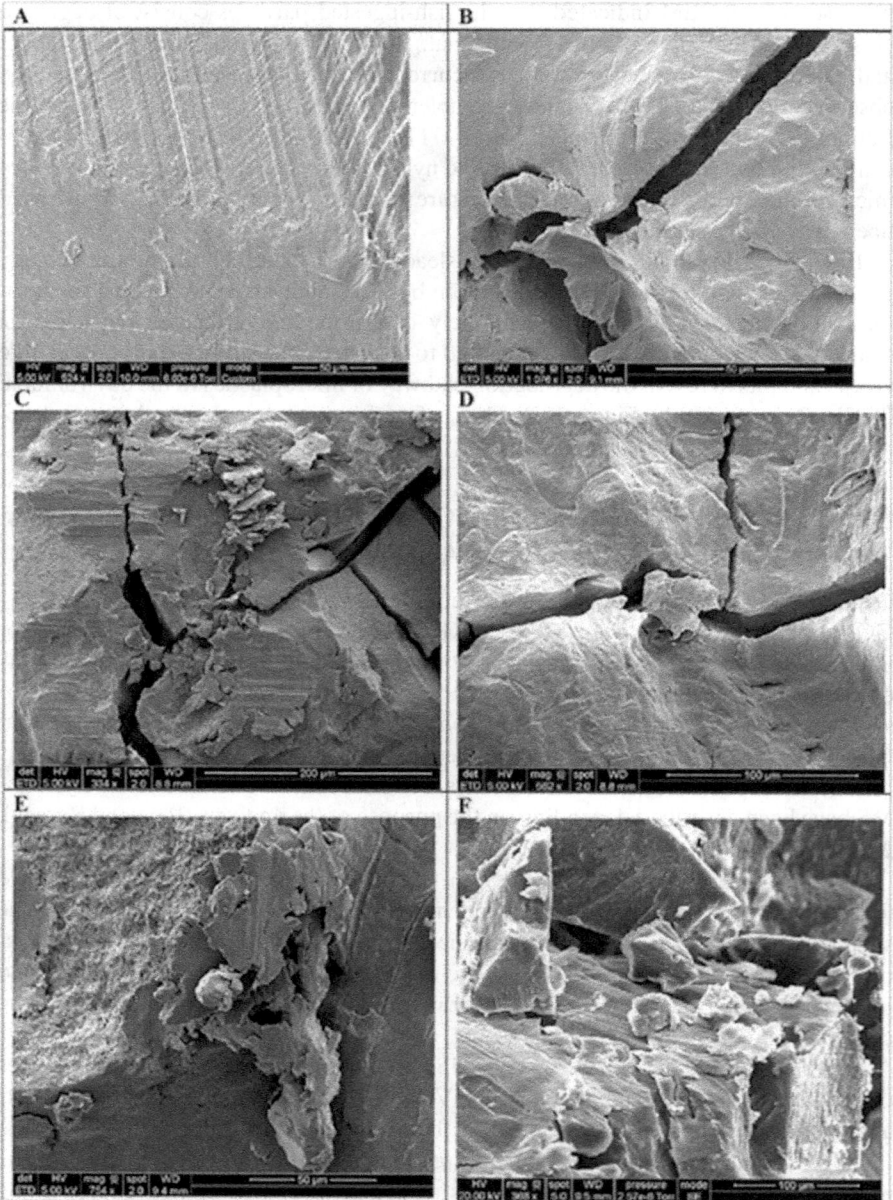

FIGURE 8.3 SEM images of a virgin pellet of high-density PE (a), and of 5 microplastics in PE previously washed with 1 M sodium hydroxide solution (b–f). (Reprinted from Ter Halle et al., 2017 after permission.)

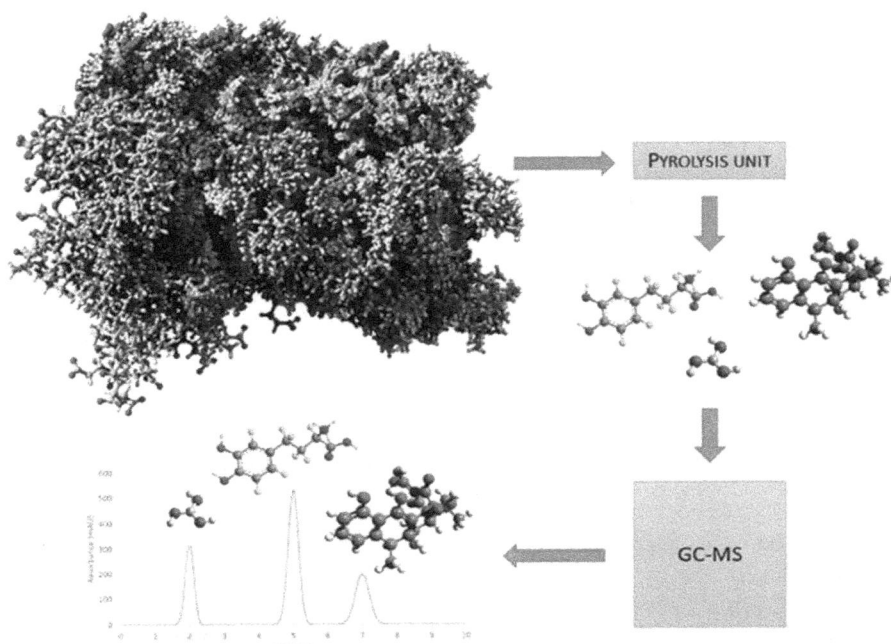

FIGURE 8.4 Fundamentals of Py–GC–MS. (Adopted from Picó and Barceló, 2020 after permission.)

macromolecules degradation that takes place in the pyrolyzer using a slow temperature ramp (instead of reaching the pyrolyzation temperature as quickly as possible) and the replacement of the chromatographic column by a short and narrow (2.5 m, 0.15 mm i.d.) deactivated capillary tube without a stationary phase to connect GC-injector and MS-detector directly. This is very similar to thermogravimetric analysis. The weight loss of the sample cannot be measured, but the result shows the thermogram, such as a differential thermogravimetric curve for a given sample. EGA–MS is done as the previous step in order to identify the temperature range for the components and set up Py–GC–MS for a more in-depth study of the identified compound.

Heart-cut analysis (Heart-cut EGA-GCMS) is the two-dimensional way of working in the combination of Py–GC. In this technique, EGA is used to obtain a thermogram, and each temperature zone of interest is analyzed separately by heart cutting evaporated components and selectively introducing them to a GC column where they are temporarily trapped at the beginning of the column prior to analyzing them by GC–MS. This technique allows examination of regions that may contain multiple components under a peak but are not obvious during Py–GC–MS analysis. This method could be incredibly useful to search both for specific components in a highly complex matrix or the whole composition of a complex system.

Reactive Py–GC–MS, the macromolecule undergoes a chemical derivatization reaction in the pyrolysis chamber that may or may not be supplied additional heat.

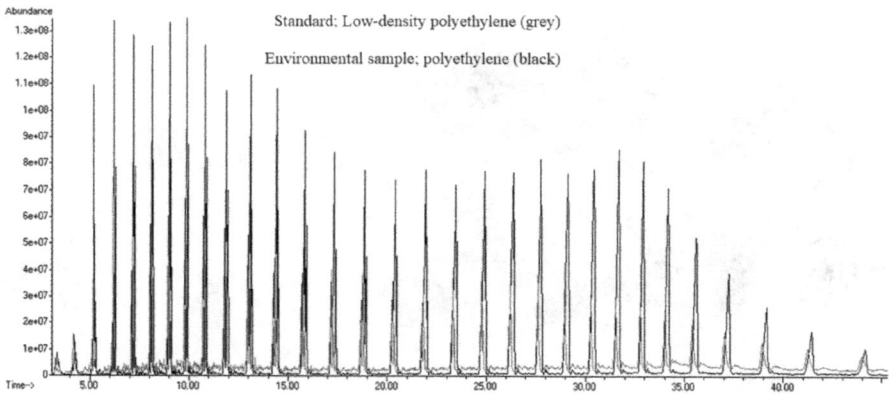

FIGURE 8.5 Pyrogram of PE (black) found in the environmental microplastic overlaid by the pyrogram of a low-density polyethylene (grey) standard. (Adapted from Fries et al., 2013 under Creative Commons 3.0 license.)

The most used derivatized agent is tetramethylammonium hydroxide (TMAH) (Picó and Barceló, 2020).

It diminishes the need for pretreatment due to a direct examination of solid polymers, in addition to a small quantity of sample analysis possible in one measurement (i.e., 5–200 µg) (Kusch, 2017). Figure 8.5 depicts a Py–GC–MS chromatogram (or say pyrograms) of PE from an environmental sample of microplastic overlaid by a standard PE (Fries et al., 2013).

A Py–GC–MS is also used for the simultaneous detection of polymer types and organic plastic additives (Fries et al., 2013). From a marine sediment sample, the particles of PP, (chlorinated)PE, PA, PS, and chlorosulfonated PE have been identified together with the polymers of dibutyl phthalate, diethylhexyl phthalate, diisobutyl phthalate, diethyl phthalate, benzaldehyde, dimethyl phthalate, and 2,4-di-*tert*-butylphenol (Fries et al., 2013).

In a study by McCormick et al. (2016), the Py–GC–MS served only as an additional technique to characterize suspected microplastics that were sorted by visual techniques. Fibers, pellets, and fragments were the common microplastics found in the sample, which contained low-density PS, PE, PP, and ethylene/propylene rubber (ethylene propylene diene monomer). For testing digestion protocols' impact on the integrity of the known sample, Dehaut et al. (2016) found it reliable to use Py–GC–MS to identify the polymer type, albeit to establish the differences of polymer subtypes was not possible. Py–GC–MS does not allow to observe the type of sample morphology, as it only gives the polymer mass per sample (Hanvey et al., 2017), thus needing a pre-selection of microplastics using optical techniques (McCormick et al., 2016; Fries et al., 2013). This leads to the use of Py–GC–MS solely as a strategy for verifying the composition of suspected microplastics (McCormick et al., 2016; Dehaut et al., 2016). In some cases, the small sample requirement may be advantageous, though it may compromise the representativeness of sample composition with the inhomogeneous sample when the complex environmental samples are analyzed (Dümichen et al., 2015; Dümichen et al., 2017). Henceforth, various new methods developed viz., the thermo-extraction and desorption integrated with a GC–MS (TED–GC–MS) (Dümichen et al., 2015; Dümichen et al., 2017). A TED–GC–MS combining thermogravimetric analysis and thermal desorption gas chromatography–mass spectrometry (TD–GC–MS) allows rapid quantification of microplastics of the polymers types (PS, PP, PE, polyamide 6, and PET) in environmental samples (Dümichen et al., 2015). Figure 8.6 depicts the result obtained from a TED–GC–MS analysis, in which the fragmented ion ($m/z = 55$) was chosen due to the presence in all aliphatic compounds with high response (Dümichen et al., 2015). Usually, the pyrograms show numerous groups with 3–5 peaks (Figure 8.6).

For a PE, among the triple peaks, the second peak exhibited the highest response in each group. The first peaks of the groups were assigned to dialkenes with two double bonds at the ends; the second peak was identified as monoalkene with one double bond mainly at one end, and the third was considered as a saturated alkane (Dümichen et al., 2015). Nonetheless, information about plastic's size and morphology is lost; Py–GC–MS gives fast measurements that can be useful for routine analyses (Dümichen et al., 2017). The method applied to samples obtained from rivers and a biogas plant identified PE, PP, and PS from the biogas plant and PS and PE (Dümichen et al., 2017). In a different method developed by Fischer and Scholz-Böttcher (2017) using Curie-Point Py–GC–MS and thermo-chemolysis,

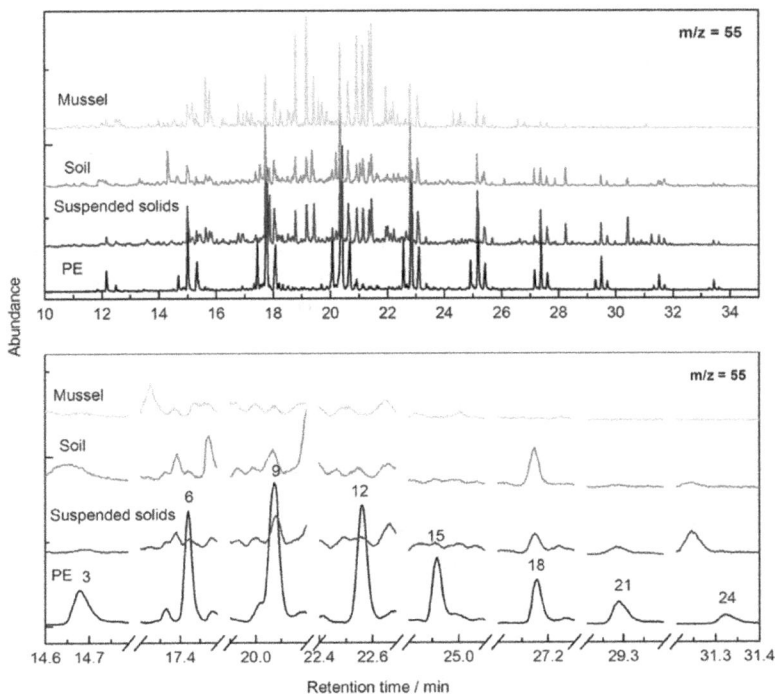

FIGURE 8.6 Overlap of the ion pyrogram of the $m/z = 55$ of PE and the environmental samples (top), and detailed view of the dialkenes of PE in comparison to the environmental samples (below). (Adopted from Dümichen et al., 2015 after permission.)

simultaneous detection of eight common polymers (PS, PP, PE, PVC, PET, poly(-methyl methacrylate), polyamide, and polycarbonate were observed. The method was also examined in fish samples spiked with known polymers, which showed a desirable quantity of microplastics (Fischer and Scholz-Böttcher, 2017).

8.3 FOURIER-TRANSFORM INFRARED SPECTROSCOPY AS AN IDENTIFICATION TOOL AND ASSOCIATED ANALYTICAL CHALLENGES

Fourier-transform infrared (FTIR) spectroscopy is a well-known and non-destructive analytical technique for identifying the types of microplastics in the samples (Harshvardhan and Jha, 2013; Titow, 1986). The technique is highly accurate in identifying specific plastic types, depending upon the distinct band patterns exhibited for a functional group. The FTIR works on an inherent light adsorption property of a molecule in the infrared region of the electromagnetic spectrum, thereby the band gaps and patterns allowing to differentiate between the plastic and natural materials.

With a wavelength longer than visible light and outside the red region of the visible spectrum, the IR-light exhibits a wavelength of $750\,nm^{-1}$. If a sample is irradiated with a beam of infrared light, analysis of the elements such as carbon, hydrogen, nitrogen, and oxygen can be undertaken by measuring the degree to which the molecules in the sample absorb specific wavelengths of the infrared light. The photons which make up the infrared light may be absorbed by the sample (absorption) or may not interact with the sample and pass straight through (transmittance). The molecules of the sample which absorb photons gain energy and as a consequence, the bonds of the molecules will distort more energetically by means of bending and stretching (Figure 8.7).

Thereby measuring the IR-absorption at different frequencies, a spectrum can be generated to underpin the molecular structure of a sample. IR-spectrum consists of a series of absorption peaks corresponding to different vibrational frequencies between the atomic bonds of sample molecules. Each plastic has different atomic combinations; hence, no two plastic materials can exhibit similar IR-spectrum. Hence, they are uniquely employed to identify the composition of a plastic sample. Due to the collection of high-resolution data between a wide frequency range of $600–4,000\,cm^{-1}$, FTIR suits to identify different functional groups of specific atoms.

However, infrared spectroscopy can be problematic in the preparation of samples. For example, the utilization of transmission techniques requires that a sample is adequately transparent so that infrared wavelengths can pass into, and transmit through, the sample. With most polymers, this is simply not attainable. Thus, suitable sample processing is required to allow the transmission of infrared radiation. This can be achieved in three ways:

1. Suspension of the polymer within a compressed potassium bromide disc that is transparent to infrared light.
2. Dispersion of the polymer in mineral oil (nujol mull).
3. Dissolution of the polymer in a solvent.

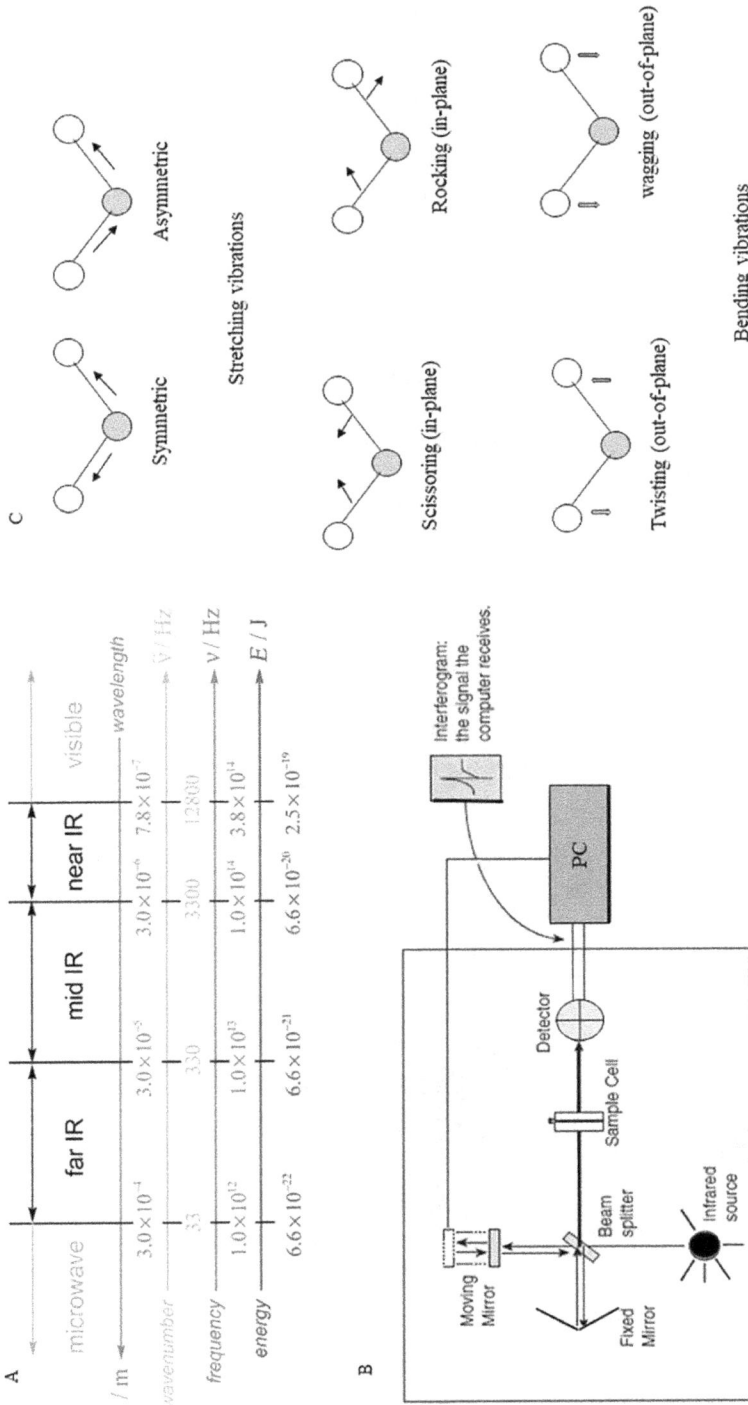

FIGURE 8.7 The infrared region of the electromagnetic spectrum (a), general schematics of FTIR (b), and (c) molecules bending and stretching vibrations when exposed to infrared radiation.

But it is hard to achieve the transparent disc requirement with potassium bromide, hence, another reflectance technique, such as attenuated total reflection (ATR) is employed.

The potassium bromide method is regarded as being predominantly difficult in that the attainment of a suitably transparent disc requires a very specific set of conditions to be met. For this reason, relatively large pieces of plastic are best analyzed using a reflectance technique, such as ATR. ATR only requires that the sample is in sufficiently close proximity to a small crystal. This crystal is typically composed of either germanium (Ge), diamond or zinc selenide (ZnSe). Once in contact with the crystal, an evanescent wave propagates 0.5–5 μm past the crystal surface and into the surface of the sample. Thus, the main requirement with ATR is that the sample is sufficiently in contact with this crystal to allow penetration of infrared radiation into the sample. This can be achieved by making use of a clamp to press the sample against the crystal (Figure 8.8).

Nevertheless, ATR is not a completely flawless technique and does exhibit some difficulties. For example, the material being analyzed must have a refraction index which is lower than the crystal, otherwise infrared light will be lost to the sample. Furthermore, the degree to which a sample is in contact with the crystal directly affects the intensity of the observed bands. This is because shorter wavelengths cannot penetrate as deeply into the sample. Consequently, the effect on intensity is

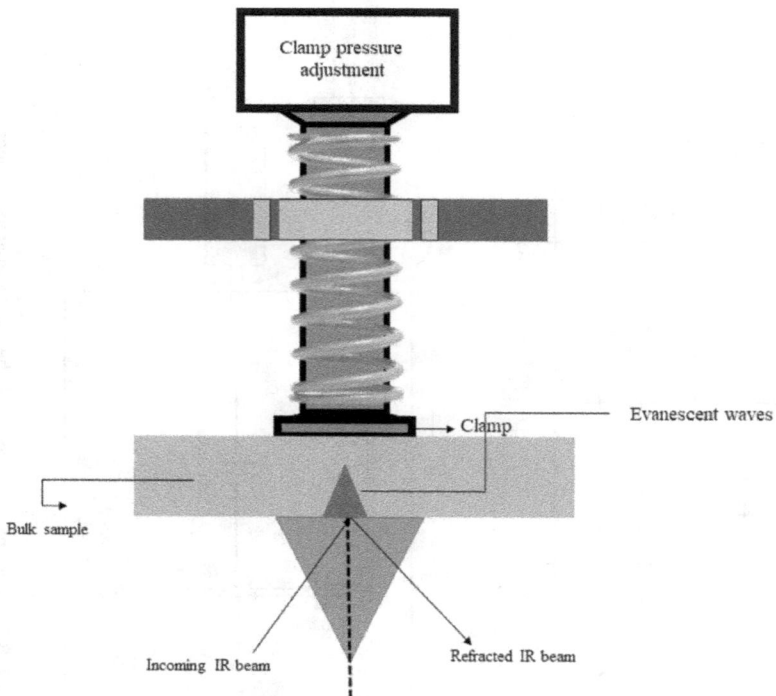

FIGURE 8.8   Sketch of attenuated total reflection.

greatest between 2,800 and 4,000 cm^{-1}. Importantly, signals from C–H, O–H, and N–H vibrations are represented in this region and, consequently, may go unnoticed if insufficient pressure is applied. Furthermore, it is important to apply sufficient pressure to prevent air from becoming trapped between the sample and the crystal, thereby ensuring that the evanescent wave will travel through the sample and not any trapped air. For this reason, the instrumental software typically displays a pressure gauge on the user interface, which provides feedback as to the amount of clamp pressure being applied. The clamp can then be adjusted until the indicated pressure is within the recommended range to produce an adequate spectrum. However, handling and clamping microplastics that are smaller than 500 μm can be difficult, and consequently, the resultant spectra are not always reliable for microplastics of this size (Wheatley et al., 1993).

Thus, for microplastics below a 500 μm size, an FTIR can be used in various modes of transmittance, reflectance, or ATR, allowing the spectrum collection and mapping of the sample. Incidentally, ATR–FTIR microscopy is advantageous over a simple FTIR technique in reflectance mode as it is able to identify the absorbance between 750 and 700 cm^{-1} corresponding to the scissoring of C–H bonding. However, minute microfibers can be lost while preparing the sample to analyze through the ATR–FTIR microscopy (Wheatley et al., 1993). Incidentally, the fingerprint region (from 1,450 to 600 cm^{-1}) is difficult to assign the absorption bands due to the unique nature of the peaks in this region. Therefore, identifying the scissoring of C–H bonds by ATR–FTIR can be useful to distinguish different types of microplastics, like PE and PVC. In a reflectance mode, FTIR is able to molecular mapping for detecting microplastics in sediments without visual identification; however, it is extremely difficult with irregular-shape microplastics due to the refractive errors, and must be used in ATR mode. However, ATR mode requires contact of a crystal (e.g., diamond) with the sample surface. Therefore, the samples and crystal surfaces are prone to contamination, and samples with hard surfaces can damage the crystal. Furthermore, focusing on every particle on a filter via a micro-ATR unit is very time consuming.

More recently, FTIR microscopes equipped with a focal plane array (FPA) detector have been assessed for the identification of microplastics. The FPA detector allows a large sample area to be measured with a high degree of lateral resolution in a short amount of time, typically within minutes. The FPA detector typically consists of a 128 × 128 array of pixels, with each pixel capable of producing an individual spectrum. Thus, over 16,000 individual spectrums of the entire sample area can be rapidly acquired simultaneously.

Tagg et al. (2015) employed (FPA)-FTIR micro-spectroscopy to detect different microplastic samples (like PE, PP, PVC, nylon-6, and PS of 150–250 μm size) from the effluent generated by wastewater treatment via H_2O_2-pretreatment (Tagg et al., 2015). A considerably reduced time with sample imaging time below 9 h was observed (Tagg et al., 2015). Recently, ATR–FTIR was used to determine polymer particles from surface water and sediments, revealing the presence of PP (51%), low-density PE (18%), high-density PE (26%), a blend of PPEP (4%), and styrene-acrylonitrile (1%) (Tsang et al., 2017). In another study, the microplastics in atmospheric fallout samples showed the presence of fibers, pellets, fragments, and films through a visual inspection followed by a digital microscope, identifying PE (14%), PS (4%), PP (9%), and cellulose (73%) (Cai et al., 2017).

As such, FTIR microscopy is also considered to be an imaging technique that provides information about the molecular composition of the sample area. However, FTIR imaging is performed in transmission mode, and due to the attributes of a suitable filter, the FTIR microscope has generally only been capable of imaging microplastics in a limited spatial frequency range (3,800–1,250 cm^{-1}), thereby greatly restricting polymer identification. However, this restriction was overcome recently with the use of a novel silicon filter substrate which possessed sufficient transparency to mid-infrared spatial frequencies in the range (4,000–600 cm^{-1}), thereby allowing successful analysis of samples for microplastics. Importantly, when reporting the use of this technique, the number of transmission mode scans and the resolution should be quoted.

Once an IR spectrum has been obtained, it can be compared with an electronic database of reference spectrums to identify the type of plastic. However, there will often be minor differences between different spectra for the same type of plastic, typically the result of impurities in the plastic material or the scanning of a sample that may not be sufficiently free from water. Thus, samples need to be sufficiently dry. Otherwise, a very large O–H stretching signal will be apparent in the spectra at around 3,300 cm^{-1}, which may mask other peaks. Furthermore, a carbon dioxide (CO_2) adsorption signal may occasionally be present at around 2,200–2,400 cm^{-1}, as a result of atmospheric CO_2. As such; it is important to run a background scan with no sample present that the instrument can subtract any atmospheric water and CO_2 signals. Additional interferences can arise from weathered plastics, which may result in poor quality signals in the spectrum and complicate identification by electronic databases. A manual inspection of the obtained spectrum is highly recommended to know about the functional groups of the plastic sample as often, there are mismatches with spectrum analysis using electronic databases due to biofouling, laboratory contamination, and surface degradation of natural organic or inorganic substances. The spectra below 60% similarity should ideally be rejected.

The microplastics can also be differentiated by irradiation in the short-wavelength infrared and near-infrared region (750–3,000 nm) of the electromagnetic spectrum. By exposing the suspected plastic to near-infrared light, the electromagnetic radiation gets adsorbed to constituent molecules to produce molecular overtone and combination vibrations. Consequently, the polymer substances can be detected by differences in the characteristic bands of C–H, N–H, and C–O. Although near-infrared is advantageous as it penetrates far deeper the plastics than FTIR, the technique is not particularly sensitive, albeit useful to examine the bulk samples.

8.4   RAMAN SPECTROSCOPY AS AN IDENTIFICATION TOOL AND ASSOCIATED ANALYTICAL CHALLENGES

The potential use of Raman spectroscopy in plastic (polystyrene) analysis was first carried out in which the detection of microplastic is conducted by directing an

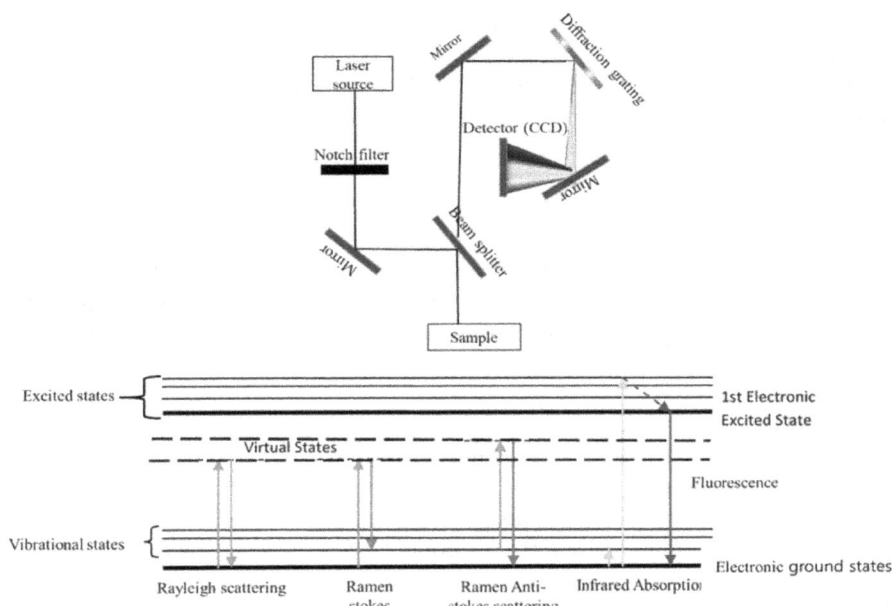

FIGURE 8.9 Schematics of Raman spectroscopy and Jablonski diagram.

incident beam of monochromatic laser light on the sample in question, which results in some of the light becoming absorbed, reflected or scattered (Figure 8.9).

The scattered light is of main interest in the Raman spectroscopy obtained as a result of photons interacting with plastic molecules, which can be divided into two forms Rayleigh scattering (elastic) and weaker Raman scattering (inelastic). Commonly, the scattered light comes under Rayleigh scattering with the same frequency as incident light; whereas, only one photon in a 30 million belongs to inelastically scattered. Thus, the small amount of inelastically scattered light produced fluorescence and mask the sample. In microplastics analysis, surface alterations ranging from sorption of humic substances to surface oxidation (aging) or biofouling are a significant source of fluorescence. Also, artificial dyes in colored plastic particles can impede detection and may result in the generation of spectra of dyes instead of polymers.

The number of inelastically scattered photons by this technique is proportional to the molecular bonds' size. So compared to Fourier transform infrared spectroscopy, water tends to have a negligible effect on the technique since the very small bonds present in a water molecule will scatter very few photons. Depending on the frequency of the incident beam, Raman inelastically scattered light can significantly increase frequency toward the blue-end of the electromagnetic spectrum through an anti-Stokes shift, or a decreased frequency toward the red-end through a Stokes shift. Such an alteration in frequency gives the vibrational frequency of a molecular bond. The Raman scattering can be described through the classical wave interpretation; the quantum particle interpretation gives a further explanation by determining the electronic states of a molecule and transitions between them. Most scattered light is elastically scattered where a photon in the incident beam is absorbed and excites a molecule in the sample to a virtual energy state. The molecule instantly relaxes back to the ground vibrational state and emits a photon of the same energy as the photon in the incident beam (Rayleigh elastic scattering). However, the molecule occasionally relaxes to a higher vibrational energy level than the ground vibrational state. The emitted photon will have less energy than the photon in the incident beam (Stokes shifted Raman scattering). Alternatively, the molecule is already in a higher vibrational energy level. The emitted photon will have more energy than the photon in the incident beam (anti-Stokes shifted Raman scattering). However, anti-Stokes scattering is less likely to occur than Stokes scattering since most molecules will tend to be in the ground vibrational state at room temperature (Smith and Dent, 2005).

A Raman spectrum is generated with characteristic peaks at associated wavelengths with different vibrations and bonds by acquiring the scattered light through a detector perpendicular to incident beams. Then, the acquired spectrum is interpreted and compared to the reference of Raman spectrums to identify plastic type. Importantly, since Raman is a surface technique, heavily fouled microplastics may need some element of clean-up prior to obtaining a spectrum (Smith and Dent, 2005).

Nevertheless, the plastic analysis particularly suits Raman due to the inherent sensitivity to non-polar molecular species. Thus, the technique readily detects the carbon bonds as C–C and C=C comprise of plastic backbone. Indeed, the technique is considered to be complementary to FTIR. Raman spectroscopy allows the observation in the polymer molecular conformity as follows:

1. The degree of crystalline regions, concerning amorphous regions, and
2. The stereo-regularity of the polymer.

Hence, Raman is established as a preferential choice to analyze the polymer morphology and gives useful information about the orientation effects and the crystalline structure of the suspected sample (Nagata et al., 1997).

8.5 NUCLEAR MAGNETIC RESONANCE SPECTROSCOPY AS AN IDENTIFICATION TOOL AND ASSOCIATED ANALYTICAL CHALLENGES

Nuclear magnetic resonance (NMR) spectroscopy is a non-destructive analytical technique used to obtain structural information about molecules.

It is based on the physical phenomenon of magnetic resonance that Isidor Isaac Rabi first demonstrated in 1938. In the 1940s, two research groups independently obtained the first successful measurements of NMR in condensed matter. The two principal investigators of these groups, Felix Bloch from Stanford University and Edward M. Purcell from Harvard University, were jointly awarded the Nobel Prize in Physics in 1952 for their contributions to magnetic resonance (Rabi et al., 1939; Purcell et al., 1946).

TABLE 8.1

Various Nuclear Spin Quantum Numbers

Atomic Number (p)	Mass Number ($p+n$)	Spin Quantum No. (I)	Examples
Even	Even	Zero	^{4}He (0), ^{12}C (0), ^{16}O (0),
Odd	Even	Integer	^{2}D (1), ^{10}B (3), ^{14}N (1)
Even or odd	Odd	Half-integer	^{15}N (1/2), ^{17}O (5/2)

The principle behind NMR is that many nuclei have a spin, and all nuclei are electrically charged. When atomic nuclei interact with externally applied radiofrequency radiations, there is a net exchange of energy which leads to a change in an intrinsic property of the atomic nuclei called nuclear spin.

The nuclear spin is defined by a quantic number (I), related to its atomic and mass number (Table 8.1), and varies depending on the considered isotope.

Only atomic nuclei with $I \neq 0$ are detectable by NMR spectroscopy (NMR-active nuclei, such as ^{1}H, ^{2}H, ^{13}C, and ^{15}N). These NMR-active nuclei behave as tiny magnets (magnetic dipoles), capable of aligning with external magnetic fields (a process called magnetization). The force of those tiny magnets is defined by a constant known as the magnetogyric ratio (γ), whose value depends on the isotope (Friebolin & Becconsall, 2005; Hore, 2015).

Nuclear spins of some NMR-active nuclei can adopt two different orientations when they align to an external magnetic field (B_0). One orientation corresponds to the lowest energy level of the nucleus (parallel to the external magnetic field), and the other one is associated with the highest energy level of the nucleus (antiparallel to the external magnetic field). The difference between energy levels (ΔE) depends on the magnetic field and the magnetogyric ratio and affects the sensitivity of the technique (Friebolin & Becconsall, 2005; Hore, 2015).

Magnetic resonance is achieved when nuclei are irradiated with radiofrequency. This causes transitions between energy levels, which involve changes in the orientation of nuclear spins.

When atomic nuclei are under the effect of a magnetic field, nuclear magnetic dipoles are not statically aligned with applied magnetic field B_0 but rather move like a spinning top (precession movement) around an axis parallel to the direction of the field. The frequency of this precession movement, called Larmor frequency (ν_L), is defined by the magnetogyric ratio and the magnetic field (Friebolin & Becconsall, 2005; Hore, 2015).

As a consequence of this precession movement, the magnetic vector (μ) associated with the nuclear magnetic dipoles possesses a component parallel to the magnetic field (μ_z) and another component perpendicular to the magnetic field (μ_{xy}), with this last one having a net value of zero in the absence of external perturbations. Notably, in an NMR experiment, it is not possible to measure the signal in the z-direction, as the magnetic field is too intense in that direction. Therefore, it is necessary to transfer the magnetization of the z component to the XY plane. For this purpose, a magnetic pulse containing frequencies close to the Larmor frequency is applied perpendicular to B_0 to reach the resonance of nuclear spins, which generates a non-zero

μ_{xy} component. After this pulse, a relaxation process occurs, and the μ_{xy} component gradually recovers its net value of zero. As a consequence of this relaxation, energy is emitted as radiofrequency, producing a characteristic signal called free induction decay which is registered by the detector. This free induction decay is subsequently transformed into a plot of intensities versus frequencies known as an NMR spectrum (Friebolin & Becconsall, 2005; Hore, 2015).

Mostly NMR spectrometers consist of three main components: a superconducting magnet, a probe, and a complex electronic system (console) controlled by a workstation (Figure 8.10).

The magnet is responsible for generating a strong magnetic field that aligns the nuclear spins of the atoms present in the sample. Nowadays, the magnets used in NMR spectroscopy are based on superconducting materials, and thus, they require very low temperatures to work (around 4 K). For this reason, NMR spectrometers contain a cooling system composed of an inner jacket filled with liquid helium, which is refrigerated by an additional jacket filled with liquid nitrogen, and many layers of thermal isolating materials (Friebolin & Becconsall, 2005; Derome, 2013).

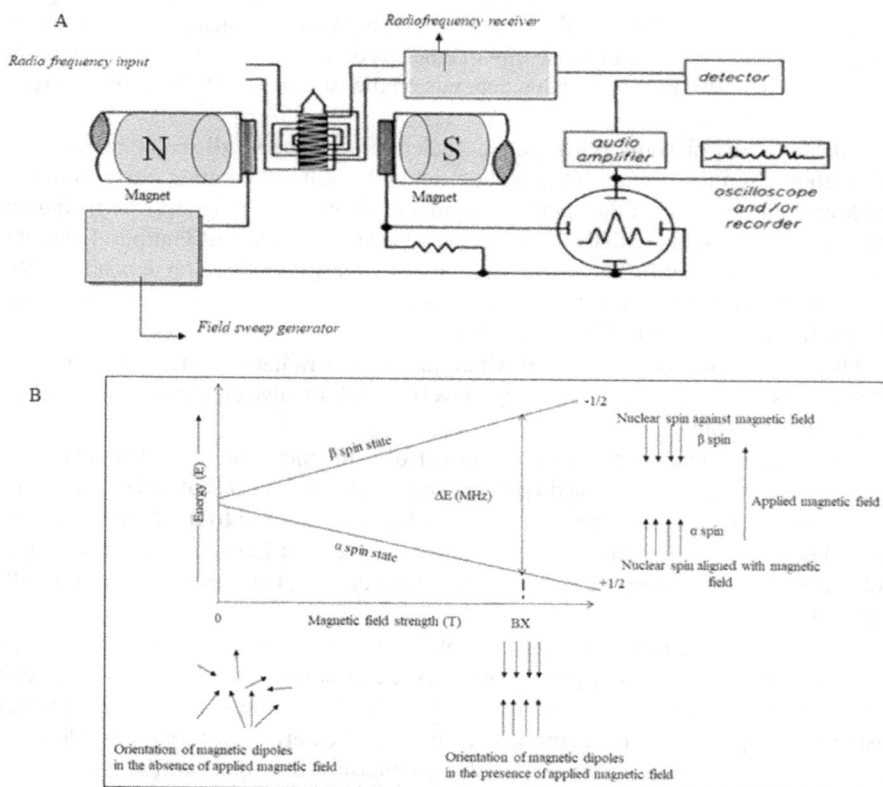

FIGURE 8.10 Schematics of NMR spectrometer (a) and orientation of magnetic dipoles in the absence and presence of magnetic field and two energy levels of nucleus as a function of applied magnetic field (b).

The superconducting magnet surrounds a cylindrical chamber known as the "probe," a crucial component of the instrument. The sample is introduced into the probe and thus placed under the influence of the magnetic field. Additionally, the probe contains a series of magnetic coils that are also located around the sample. These coils have multiple purposes. On the one hand, they are used to irradiate the radiofrequency pulses and detect and collect the NMR signal emitted by the sample. On the other hand, they also enable control of the magnetic field homogeneity and the application of pulse gradients that are used in some NMR experiments (Friebolin & Becconsall, 2005; Derome, 2013).

Finally, the electronic system of the spectrometer controls all the experimental conditions and enables the setup and modification of every parameter of the NMR experiment through the workstation. This system is also responsible for data acquisition and subsequent mathematical transformation into an NMR spectrum. The spectrum contains a series of peaks of different intensities as a function of a magnitude known as the chemical shift that is derived from the Larmor frequency of the different atomic nuclei present in the sample (Friebolin & Becconsall, 2005; Derome, 2013).

The signal detected by an NMR spectrometer (the free induction decay) must be transformed prior to analysis. As the Larmor frequency is dependent upon the intensity of the magnetic field, it varies from instrument to instrument. For this reason, a mathematical transformation is performed to provide a relative magnitude called chemical shift (δ). Unlike the Larmor frequency, this magnitude is independent of the magnetic field, and the value can be compared across instruments (Friebolin & Becconsall, 2005; Derome, 2013; Hore, 2015). Here, ν_L is the observed Larmor frequency of a nucleus and ν_L^0 is the Larmor frequency of a reference nucleus, both in Hz. By convention, chemical shift is always expressed in parts per million (ppm). The zero value of the chemical shift scale is set using a reference compound (such as tetramethylsilane or sodium trimethylsilylpropanesulfonate for ^{1}H).

An NMR spectrum provides a lot of information about the molecules present in the sample. First, chemical shift values can identify chemical groups within a molecule. For example, acetic acid ($H_3C-COOH$) has four protons, and the three protons of the methyl group (CH_3) are magnetically equivalent and therefore have the same chemical shift. So one signal comes from the CH_3 group and the other from the proton in the carboxylic acid group (COOH). Secondly, in ^{1}H-NMR spectra, the signal area is proportional to the number of atomic nuclei producing that signal (this does not apply to ^{13}C-NMR spectra). In this example, if the areas of both signals were to be calculated, the most intense signal would be three times larger than the other. It is by the fact that one signal represents the three protons from the CH_3 group (signal at $\delta = 2.0$ ppm) and the other one the proton from the COOH group (signal at $\delta = 11.5$ ppm) (Jacobsen, 2016; Bible, 2013).

The spins of two nuclei connected through a few chemical bonds can interact, causing a phenomenon known as scalar coupling that splits the signals. Typically, this coupling is only observable when the number of chemical bonds separating two nuclei does not exceed four. The splitting of the signals follows a pattern that depends on the number of coupled nuclei and a coupling constant (J) defined by the type of nuclei and the distance (in chemical bonds) between them. The characteristic shape of a split signal is called multiplicity and provides additional information about the molecule. This multiplicity can be calculated using the $N+1$ rule. This rule states that if a proton shows scalar coupling with N protons attached to contiguous carbon nuclei, its signal will split into $N+1$ peaks with relative intensities defined by Pascal's triangle. Peak splitting because of the scalar coupling causes a reduction of the peak intensity. Finally, the observation of signals arising from an effect called the nuclear Overhauser effect is essential for the structural determination of macromolecules since it emerges from the interaction of nuclear spins of atoms that are spatially close but distant in the molecular sequence (Friebolin & Becconsall, 2005; Derome, 2013; Hore, 2015; Jacobsen, 2016; Bible, 2013).

In this context, to interpret an NMR spectrum, it is necessary to use all that information to assign each observed signal to the corresponding atomic nucleus of the molecule(s) in the sample. This process is called a spectral assignment, and it can be difficult to achieve with complex molecules. For this reason, many types of NMR experiments providing different and complementary information are used to characterize a sample (Wüthrich, 1990).

The same kind of nuclei can generate signals with different chemical shift values. These chemical shifts differ as the magnetic field sensed by a particular nucleus strongly depends on its local chemical environment. The circulation of electrons in the surroundings of a nucleus creates small magnetic fields that oppose the applied external field. This "shielding" effect (σ) is directly proportional to the electronic density around the nucleus. As a result, the effective magnetic field acting on the nucleus is lower, and the Larmor frequency is affected. When there is a high electronic density around the considered nucleus, the shielding effect is high, the Larmor frequency decreases, and so does the chemical shift (it moves upfield). On the contrary, when the electronic density is low in the vicinity of the nucleus, the shielding effect is low, the Larmor frequency takes higher values, and so does the chemical shift (it moves downfield). Therefore, in NMR spectroscopy, upfield and downfield refer to the regions of lower and higher values, respectively, within the chemical shift scale (Friebolin & Becconsall, 2005; Derome, 2013; Hore, 2015).

Hydrogen nuclei from methyl groups or aliphatic molecules are strongly shielded, and their typical chemical shift values are located upfield. On the other hand, hydrogen nuclei attached to electronegative atoms (such as oxygen or nitrogen) or close to electronegative groups (such as carboxylic acids or aldehydes) are deshielded and show chemical values located downfield. By convention, the chemical shift scale in an NMR spectrum is represented from right to left. As described above, the zero value is established using a standard compound whose carbon and hydrogen atoms are strongly shielded, and hence, their signals appear in the furthest upfield region. The assignment of the NMR spectra is usually performed with the help of NMR charts or diagrams that facilitate the identification of the NMR signals.

Hydrogens or carbons that are highly shielded, such as methyl groups, have low chemical shift values. However, hydrogens attached to very electronegative groups (e.g., carboxylic acids, ketones, or aldehydes) have high chemical shift values.

These charts represent typical chemical shifts, but sometimes the values could be displaced to other regions of the scale (Du Vernet and Boekelheide, 1974). For instance, in large macromolecules, a distant chemical group can be relocated due to spatial rearrangements of the tridimensional structure. This relocation could alter the chemical environment of the measured nucleus, leading to a change in its chemical shift value.

Notably, the natural abundance of an NMR-active atomic nucleus and the gyromagnetic ratio (measured between the nuclear magnetic moment and spin angular momentum denoted by γ) directly affect the sensitivity to that nucleus. Hence, the abundant nuclei with a high γ-value (e.g., ^{1}H and ^{19}F) are the most responsive to the NMR technique.

Peez and Imhof (2020) analyze polyvinyl chloride (PVC) powder with a size <50 μm, acrylonitrile butadiene styrene (ABS) granules with a size distribution of 100–300 μm, and polyamide (PA) fibers with a length of approx. 500 μm and a diameter of approx. 20–30 μm with ^{1}H-NMR (qNMR) spectroscopy. For quantification, the integration method or the peak-fitting method combined with the calibration curve method was used. Linearity above $0.99R^2$, the precision of 99.1%–99.9%, and the accuracy of 95.6%–110% for all three polymer types demonstrate the high analytical potential of the method. Moreover, the detection limit (40–84 μg mL^{-1}) for all polymer types is in the lower environmentally relevant range.

Overall, the technique is very useful in the determination of the chemical structure of the polymer chain of plastic materials and highly detailed information can be obtained relating to the sequence of monomers in a copolymer and the degree of crystallinity in a semi-crystalline plastic, as well as information relating to branching and tacticity. For example, the stereochemistry of polypropylene can be examined to ascertain the tactic form present. Furthermore, chemical changes in the plastic material can also be detected, such as oxidation states, with a high degree of sensitivity but further efforts are required to establish standard protocols for efficient analysis of various plastics.

8.6 MALDI–TOF MASS SPECTROMETRY AS AN IDENTIFICATION TOOL AND ASSOCIATED ANALYTICAL CHALLENGES

Matrix-assisted laser desorption/ionization–time-of-flight mass spectrometry (MALDI–TOF MS) is a powerful identification technique for ionizing and detecting intact molecules with high molecular weights. It consists of three essential components: ionization, the time of flight phase (separation), and detection. The samples are pipetted onto a stainless steel sample plate and then vaporized and ionized into the ionization components. Then, the charged analytes will be separated under different voltages and detected by mass spectrometry based on their mass-to-charge ratio (m/z). With advantages like a simple operation, high sensitivity, high-throughput analysis, and reliable results, the MALDI–TOF MS method has attracted increasing attention in the identification and quantification, as well as the distribution of emerging pollutants in environmental and biological samples. Using this technique, Yang et al. examined the ecotoxicity of perfluorooctanesulfonic acid in mouse kidney tissues (Payne and Grayson, 2018). Previous studies reported the possibility of analyzing pure polymers by the MALDI–TOF MS technique (Payne and Grayson, 2018) and showed the formation of [polymer]metal+ ions by mixing the polymer with metal cationization agents like sodium, potassium, and silver salts. Microplastics and nanoplastics are mainly composed of various polymers. It is hypothesized that the MALDI–TOF MS technique can be further employed to detect environmental microplastics and nanoplastics. Once the polymers were softly ionized and carried by the vaporized matrix after absorbing the laser energy, they would enter the TOF and be separated according to their m/z. Finally, the MS results could be obtained for further analysis, including the repeating unit mass, end groups, and molecular formula, which are the most critical information for identifying microplastics and nanoplastics, especially after aging or degradation processes (Wu et al., 2020).

Wu et al. (2020) further investigated by using MALDI–TOF MS that PS-related plastics (micro-, nano-) (fresh plastics and the aged) consisted of C_8H_8 and $C_{16}H_{16}O$ oligomers, while the PET-related plastics (micro-, nano-) (only found in sediment) were identified with repeated units of $C_{10}H_8O_4$ and $C_{12}H_{12}O_4$. According to the quantitative correlation curve, the contents of PS and PET microplastics and nanoplastics were quantified as 8.56 ± 0.04 and 28.71 ± 0.20 mg·kg^{-1}, respectively, in the collected sediment.

Dimzon et al. (2012) and Weidner and Trimpin (2010) examined the suitability of MALDI–TOF MS to detect various polymers with different sample preparation techniques. A limitation is given by the fact that different polymers require adapted cationizing agents for their ionization. However, the technique cannot supply the size, shape, and distribution of the microplastics and nanoplastics directly at present. MALDI–TOF spectra can provide morphological information such as the size and shape of particles through imaging techniques. Rivas et al. (2016) used MALDI–TOF mass spectrometry imaging to examine modifications of polymer surfaces during degradation. Samples were placed on indium-tin-oxide glass slides, and the matrix was sublimated and deposited into the target through a special coating chamber. This technique enables the determination of spatial changes of a polymer surface.

ACKNOWLEDGMENTS

This work was supported by a grant from the National Research Foundation of Korea (NRF) grant funded by the Korea government (MSIT) (No. NRF-2020R1A2C1013851).

REFERENCES

Cai, L., Wang, J., Peng, J., Tan, Z., Zhan, Z., Tan, X., Chen, Q. (2017) Characteristic of microplastics in the atmospheric fallout from Dongguan city, China: preliminary research and first evidence. *Environmental Science and Pollution Research*, 24(32), 24928–24935.

Dehaut, A., Cassone, A. L., Frère, L., Hermabessiere, L., Himber, C., Rinnert, E., Paul-Pont, I. (2016) Microplastics in seafood: Benchmark protocol for their extraction and characterization. *Environmental Pollution*, 215, 223–233.

Dehghani, S., Moore, F., Akhbarizadeh, R. (2017) Microplastic pollution in deposited urban dust, Tehran metropolis, Iran. *Environmental Science and Pollution Research*, 24(25), 20360–20371.

Du Vernet, R., Boekelheide, V. (1974) Nuclear magnetic resonance spectroscopy. Ring-current effects on carbon-13 chemical shifts. *Proceedings of the National Academy of Sciences*, 71(8), 2961–2964. doi: 10.1073/pnas.71.8.2961.

Dümichen, E., Barthel, A. K., Braun, U., Bannick, C. G., Brand, K., Jekel, M., Senz, R. (2015) Analysis of polyethylene microplastics in environmental samples, using a thermal decomposition method. *Water Research*, 85, 451–457.

Dümichen, E., Eisentraut, P., Bannick, C. G., Barthel, A. K., Senz, R., Braun, U. (2017) Fast identification of microplastics in complex environmental samples by a thermal degradation method. *Chemosphere*, 174, 572–584.

Eriksen, M., Maximenko, N., Thiel, M., Cummins, A., Lattin, G., Wilson, S., Hafner, J., Zellers, A., Rifman, S. (2013) Plastic pollution in the South Pacific subtropical gyre. *Marine Pollution Bulletin*, 68(1–2), 71–76.

Fischer, M., Scholz-Böttcher, B. M. (2017) Simultaneous trace identification and quantification of common types of microplastics in environmental samples by pyrolysis-gas chromatography–mass spectrometry. *Environmental Science & Technology*, 51(9), 5052–5060.

Friebolin, H., Becconsall, J. K. *Basic One-and Two-Dimensional NMR Spectroscopy (Vol. 7)*. Weinheim: Wiley-vch, 2005.

Fries, E., Dekiff, J. H., Willmeyer, J., Nuelle, M. T., Ebert, M., Remy, D. (2013) Identification of polymer types and additives in marine microplastic particles using pyrolysis-GC/MS and scanning electron microscopy. *Environmental Science: Processes & Impacts*, 15(-10), 1949–1956.

Hanvey, J. S., Lewis, P. J., Lavers, J. L., Crosbie, N. D., Pozo, K., Clarke, B. O. (2017) A review of analytical techniques for quantifying microplastics in sediments. *Analytical Methods*, 9(9), 1369–1383.

Harshvardhan, K., Jha, B. (2013) Biodegradation of low-density polyethylene by marine bacteria from pelagic waters, Arabian Sea, India. *Marine Pollution Bulletin*, 77(1–2), 100–106.

Hore, P. J. *Nuclear Magnetic Resonance*. Oxford: Oxford University Press, 2015.

Jacobsen, N. E. *NMR Data Interpretation Explained: Understanding 1D and 2D NMR Spectra of Organic Compounds and Natural Products*. Hoboken: John Wiley & Sons. 2016.

Kusch, P. (2017) Application of pyrolysis-gas chromatography/mass spectrometry (Py-GC/MS). *Comprehensive Analytical Chemistry*, 75, 169–207.

Li, J., Qu, X., Su, L., Zhang, W., Yang, D., Kolandhasamy, P., Shi, H. (2016) Microplastics in mussels along the coastal waters of China. *Environmental Pollution*, 214, 177–184.

McCormick, A. R., Hoellein, T. J., London, M. G., Hittie, J., Scott, J. W., Kelly, J. J. (2016) Microplastic in surface waters of urban rivers: concentration, sources, and associated bacterial assemblages. *Ecosphere*, 7(11), e01556.

Nagata, M., Kiyotsukuri, T., Minami, S., Tsutsumi, N., Sakai, W. (1997) Enzymatic degradation of poly(ethylene terephthalate) copolymers with aliphatic dicarboxylic acids and/or poly(ethylene glycol). *European Polymer Journal*, 33(10–12), 1701–1705.

Napper, I. E., Thompson, R. C. (2016) Release of synthetic microplastic plastic fibres from domestic washing machines: Effects of fabric type and washing conditions. *Marine Pollution Bulletin*, 112(1–2), 39–45.

Payne, M. E, Grayson, S. M. (2018) Characterization of synthetic polymers via matrix assisted laser desorption ionization time of flight (MALDI-TOF) mass spectrometry. *Journal of Visualized Experiments*, 136, 57174. doi: 10.3791/57174. PMID: 29939185; PMCID: PMC6101691.

Peez, N., Imhof, W. (2020) Quantitative 1 H-NMR spectroscopy as an efficient method for identification and quantification of PVC, ABS and PA microparticles. *Analyst*, 145, 5363.

Picó, Y., Barceló, D. (2020) Pyrolysis gas chromatography-mass spectrometry in environmental analysis: Focus on organic matter and microplastics. *TrAC Trends in Analytical Chemistry*, 130, 115964.

Purcell, E. M., Torrey, H. C., Pound, R. V. (1946) Resonance absorption by nuclear magnetic moments in a solid. *Physical Review*, 69, 37L. doi: 10.1103/PhysRev.69.37.

Qiu, Q., Tan, Z., Wang, J., Peng, J., Li, M., Zhan, Z. (2016) Extraction, enumeration and identification methods for monitoring microplastics in the environment. *Estuarine, Coastal and Shelf Science*, 176, 102–109.

Rabi, I. I., Millman, S., Kusch, P., Zacharias, J. R. (1939) The molecular beam resonance method for measuring nuclear magnetic moments. the magnetic moments of $_3Li^6$,$_3Li^7$ and $_9F^{19}$. *Physical Review*, 55(6), 526. doi: 10.1103/PhysRev.55.526.

Rivas, D., Ginebreda, A., Pérez, S., Quero, C., Barceló, D. (2016). MALDI-TOF MS Imaging evidences spatial differences in the degradation of solid polycaprolactone diol in water under aerobic and denitrifying conditions. *Science of the Total Environment*, 566–567, 27–33.

Silva, A. B., Bastos, A. S., Justino, C. I., da Costa, J. P., Duarte, A. C. & Rocha-Santos, T. A. (2018) Microplastics in the environment: Challenges in analytical chemistry-A review. *Analytica chimica acta*, 1017, 1–19.

Smith, W. E., Dent, G. *Modern Raman Spectroscopy: A Practical Approach*. John Wiley & Sons, Ltd. 2005. ISBN 0-471-49794-0.

Tagg, A. S., Sapp, M., Harrison, J. P., Ojeda, J. J. (2015) Identification and quantification of microplastics in wastewater using focal plane array-based reflectance micro-FT-IR imaging. *Analytical Chemistry*, 87(12), 6032–6040.

Ter Halle, A., Ladirat, L., Martignac, M., Mingotaud, A. F., Boyron, O., Perez, E. (2017) To what extent are microplastics from the open ocean weathered?. *Environmental Pollution*, 227, 167–174.

Titow, W. V. *PVC Technology*. Elsevier Applied Science Publishers Ltd. 1986. ISBN 978-94-010-8976-0.

Tsang, Y. Y., Mak, C. W., Liebich, C., Lam, S. W., Sze, E. T., Chan, K. M. (2017) Microplastic pollution in the marine waters and sediments of Hong Kong. *Marine Pollution Bulletin*, 115(1–2), 20–28.

Vandermeersch, G., Van Cauwenberghe, L., Janssen, C.R., Marques, A., Granby, K., Fait, G., Kotterman, M. J. J., Diogène, J., Bekaert, K., Robbens, J., Devriese, L. (2015) A critical view on microplastic quantification in aquatic organisms. *Environmental Research*, 143, 46–55.

Wang, Z. M., Wagner, J., Ghosal, S., Bedi, G., Wall, S. (2017) SEM/EDS and optical microscopy analyses of microplastics in ocean trawl and fish guts. *Science of the Total Environment*, 603, 616–626.

Weidner, S. M., Trimpin, S. (2010) Mass spectrometry of synthetic polymers. *Analytical Chemistry*, 82, 4811–4829.

Wheatley, L., Levendis, Y. A., Vouros, P. (1993) Exploratory-study on the combustion and Pah emissions of selected municipal waste plastics. *Environmental Science & Technology*, 27 (13):2885–2895.

Wu, P, Tang, Y, Cao, G, Li, J, Wang, S, Chang, X, Dang, M, Jin, H, Zheng, C, Cai, Z. (2020) Determination of environmental micro (nano) plastics by matrix-assisted laser desorption/ionization–time-of-flight mass spectrometry. *Analytical Chemistry*, 92(21), 14346–14356.

Wüthrich, K. (1990) Protein structure determination in solution by NMR spectroscopy. *Journal of Biological Chemistry*, 265(36), 22059–22062. doi: 10.1016/S0021-9258(18)45665-7.

Index

For Product Safety Concerns and Information please contact our EU
representative GPSR@taylorandfrancis.com
Taylor & Francis Verlag GmbH, Kaufingerstraße 24, 80331 München, Germany

www.ingramcontent.com/pod-product-compliance
Lightning Source LLC
Chambersburg PA
CBHW070713220326
41598CB00024BA/3127